T0233738

Lecture Notes in Artificial Intelligence   9653

Subseries of Lecture Notes in Computer Science

LNAI Series Editors

Randy Goebel
University of Alberta, Edmonton, Canada
Yuzuru Tanaka
Hokkaido University, Sapporo, Japan
Wolfgang Wahlster
DFKI and Saarland University, Saarbrücken, Germany

LNAI Founding Series Editor

Joerg Siekmann
DFKI and Saarland University, Saarbrücken, Germany

More information about this series at http://www.springer.com/series/1244

Alexander Gammerman · Zhiyuan Luo
Jesús Vega · Vladimir Vovk (Eds.)

Conformal and Probabilistic Prediction with Applications

5th International Symposium, COPA 2016
Madrid, Spain, April 20–22, 2016
Proceedings

Springer

Editors
Alexander Gammerman
University of London
Egham
UK

Zhiyuan Luo
University of London
Egham
UK

Jesús Vega
CIEMAT
Madrid
Spain

Vladimir Vovk
University of London
Egham
UK

ISSN 0302-9743   ISSN 1611-3349   (electronic)
Lecture Notes in Artificial Intelligence
ISBN 978-3-319-33394-6   ISBN 978-3-319-33395-3   (eBook)
DOI 10.1007/978-3-319-33395-3

Library of Congress Control Number: 2016936643

LNCS Sublibrary: SL7 – Artificial Intelligence

© Springer International Publishing Switzerland 2016
This work is subject to copyright. All rights are reserved by the Publisher, whether the whole or part of the material is concerned, specifically the rights of translation, reprinting, reuse of illustrations, recitation, broadcasting, reproduction on microfilms or in any other physical way, and transmission or information storage and retrieval, electronic adaptation, computer software, or by similar or dissimilar methodology now known or hereafter developed.
The use of general descriptive names, registered names, trademarks, service marks, etc. in this publication does not imply, even in the absence of a specific statement, that such names are exempt from the relevant protective laws and regulations and therefore free for general use.
The publisher, the authors and the editors are safe to assume that the advice and information in this book are believed to be true and accurate at the date of publication. Neither the publisher nor the authors or the editors give a warranty, express or implied, with respect to the material contained herein or for any errors or omissions that may have been made.

Printed on acid-free paper

This Springer imprint is published by Springer Nature
The registered company is Springer International Publishing AG Switzerland

Preface

This volume contains the proceedings of the 5th Symposium on Conformal and Probabilistic Prediction with Applications (COPA 2016), which was co-organized by Royal Holloway, University of London, UK, and Centro de Investigaciones Energéticas, Medioambientales y Tecnológicas (CIEMAT), Madrid, Spain. The Symposium was held at CIEMAT during April 20–22, 2016.

Conformal prediction is a recently developed framework for complementing the predictions of machine learning algorithms with reliable measures of confidence. The framework produces well-calibrated confidence measures for individual examples without assuming anything more than that the data are generated independently from the same probability distribution.

Since its development the framework has been applied to many popular techniques, such as support vector machines, k-nearest neighbors, neural networks, ridge regression etc., and has been successfully applied to many challenging real world problems, such as the early detection of ovarian cancer, the classification of leukemia subtypes, the diagnosis of acute abdominal pain, the assessment of stroke risk, the recognition of hypoxia in electroencephalograms (EEGs), the prediction of plant promoters, the prediction of network traffic demand, the estimation of effort for software projects and the back-calculation of non-linear pavement layer moduli. The framework has also been extended to additional problem settings such as semi-supervised learning, anomaly detection, feature selection, outlier detection, change detection in streams, and active learning. Recent developments in collecting large volumes of data have also required its adjustment to handle "big data".

The aim of this symposium is to serve as a forum for the presentation of new and ongoing work and the exchange of ideas between researchers on any aspect of conformal and probabilistic prediction and their applications.

While the previous four annual gatherings (COPA 2012 to COPA 2015) were devoted mainly to conformal predictors, they also included extensions of conformal predictors to Venn predictors. The title of this year's event reflects the expanded scope explicitly and covers all kinds of probabilistic prediction, not only Venn prediction.

The popularity of conformal prediction in the machine-learning community is growing. As evidence of this we can mention the following events that took place after COPA 2015. In June 2015, a special issue on "Conformal Prediction and its Applications" of the *Annals of Mathematics and Artificial Intelligence* (Volume 74, Issues 1–2) was published. In July 2015, Henrik Boström, Alexander Gammerman, Ulf Johansson, Lars Carlsson, and Henrik Linusson presented the tutorial "Conformal Prediction: A Valid Approach to Confidence Predictions" at the 2015 International Joint Conference on Neural Networks (Killarney, Ireland). An EU Horizon 2020 project on drug design that started in September 2015 adopted conformal prediction as one of the main tools for selecting useful chemical compounds. In December 2015, an Indo-UK workshop on "Mathematical Foundations of Probabilistic Conformal

Prediction and Its Applications in Machine Learning" was held at the Indian Institute of Technology in Hyderabad, India. In January 2016, there was a session on "Data-Intensive Methods and Conformal Predictions" at the International Conference on Pharmaceutical Bioinformatics (ICPB 2016) in Pattaya, Thailand.

Overall, 14 papers were accepted for presentation at the symposium after being reviewed by at least two independent academic referees. The authors of these papers come from 11 different countries, namely: Austria, Cyprus, Italy, The Netherlands, Russia, Spain, Sweden, Switzerland, Ukraine, the UK and the USA.

The volume is divided into three parts. The first part presents the invited paper "Learning with Intelligent Teacher" by Vladimir iapnik and Rauf Izmailov, devoted to learning with privileged information and emphasizing the role of the teacher in the learning process.

The second part is devoted to the theory of conformal prediction. The two papers in this part investigate various criteria of efficiency used in conformal prediction (Vladimir Vovk, Valentina Fedorova, Ilia Nouretdinov, and Alexander Gammerman) and introduce a universal probability-free version of conformal predictors (Vladimir Vovk and Dusko Pavlovic).

The core of the book is formed by the third part, containing experimental papers describing various applications of conformal prediction. This part opens by "Conformal Predictors for Compound Activity Prediction" (Paolo Toccaceli, Ilia Nouretdinov and Alexander Gammerman), applying conformal prediction to big and imbalanced datasets in the field of drug discovery. The following paper, "Conformal Prediction of Disruptions from Scratch: Application to an ITER Scenario" by Raul Moreno, Jesús Vega, and Sebastian Dormido-Canto, demonstrates advantages of conformal prediction over the conventional methodology in the field of nuclear fusion. In "Evaluation of a Variance-Based Nonconformity Measure for Regression Forests" Henrik Boström, Henrik Linusson, Tuve Löfström and Ulf Johansson continue their empirical investigation of conformal prediction based on random forests; their new algorithms achieve impressive computational efficiency while retaining predictive efficiency. This part is concluded by four papers proposing valuable extensions of the framework of conformal prediction in various directions. First, Antonis Lambrou and Harris Papadopoulos ("Binary Relevance Multi-label Conformal Predictor") extend the framework to multi-label classification. The second extension is proposed by Andrea Murari, Saeed Talebzadeh, Jesús Vega, Emmanuele Peluso, Michela Gelfusa, Michele Lungaroni, and Pasqualino Gaudio in "A Metric to Improve the Robustness of Conformal Predictors in the Presence of Error Bars": now all data, including the attributes of the objects to be labelled, are not precise but are obtained using a noisy measurement procedure. The third paper, by Shuang Zhou, Evgueni Smirnov, Ralf Peeters, and Gijs Schoenmakers ("Decision Trees for Instance Transfer"), applies the ideas of conformal prediction to the case where the test data are generated from a distribution different from that generating the training data. Finally, Giovanni Cherubin and Ilia Nouretdinov ("Hidden Markov Models with Confidence") extend the methodology of conformal prediction to the popular setting of hidden Markov models.

The third part contains theoretical and experimental papers in general machine learning. It opens by two theoretical papers, "Variable Fidelity Regression Using Low Fidelity Function Blackbox, and Sparsification" by Alexey Zaytsev and "Effective

Design for Sobol Indices Estimation Based on Polynomial Chaos Expansions" by Evgeny Burnaev, Ivan Panin, and Bruno Sudret. Apart from theoretical results, both papers provide convincing empirical validation. The next two papers are devoted to two different, both very important, applications: medicine ("Joint Prediction of Chronic Conditions Onset: Comparing Multivariate Probits with Multiclass Support Vector Machines" by Shima Ghassem Pour and Federico Girosi) and information security ("Method of Learning Malware Behavior Scripts by Sequential Pattern Mining" by Alexandra Moldavskaya, Victoria Ruvinskaya, and Evgeniy Berkovich). The final paper, "Extended Regression on Manifolds Estimation" by Alexander Kuleshov and Alexander Bernstein, solves several interrelated problems in the area of regression on manifolds.

We are very grateful to the Program and Organizing Committees; the success of the symposium would have been impossible without their hard work. We are also indebted to the sponsors: Royal Holloway, University of London, and CIEMAT. Our special thanks to Yandex for their help and support in organizing the symposium and the special Alexey Chervonenkis Memorial Lecture.

March 2016
                                                Alexander Gammerman
                                                Zhiyuan Luo
                                                Jesús Vega
                                                Vladimir Vovk

Organization

General Chairs

Alexander Gammerman   Department of Computer Science, Royal Holloway,
University of London, Egham, Surrey, UK
Vladimir Vapnik   AI Research Facebook, Columbia University, USA
and Royal Holloway, University of London, UK
Vladimir Vovk   Department of Computer Science, Royal Holloway,
University of London, Egham, Surrey, UK

Organizing Committee

Jesús Vega   Laboratorio Nacional de Fusion, CIEMAT,
Madrid, Spain
Zhiyuan Luo   Department of Computer Science, Royal Holloway,
University of London, Egham, Surrey, UK
Harris Papadopoulos   Department of Computer Science and Engineering,
Frederick University, Nicosia, Cyprus

Program Committee Chairs

Jesús Vega   Laboratorio Nacional de Fusion, CIEMAT,
Madrid, Spain
Zhiyuan Luo   Department of Computer Science, Royal Holloway,
University of London, Egham, Surrey, UK

Program Committee Members

Ernst Ahlberg   AstraZenrca, Sweden
Vineeth Balasubramanian   Department of Computer Science and Engineering,
Indian Institute of Technology, Hyderabad, India
Henrik Boström   Department of Computer and Systems Sciences,
Stockholm University, Stockholm, Sweden
Lars Carlsson   AstraZenrca, Sweden
Vladimir Cherkassky   Department of Electrical and Computer Engineering,
University of Minnesota, USA
Jesús Manuel de la Cruz   Universidad Complutense de Madrid, Madrid, Spain
Jose-Carlos Gonzalez-Cristobal   Universidad Politécnica de Madrid, Madrid, Spain
Anna Fukshansky   Mathematik – Training und Lösungen, Germany
Barbara Hammer   Bielefeld University, Bielefeld, Germany

Shenshyang Ho   School of Computer Science and Engineering, College
                                of Engineering, Nanyang Technological University,
                                Singapore
Carlo Lauro   University of Naples Federico II, Italy
David Lindsay   GKFX Financial Services, London, UK
Henrik Linusson   University of Borås, Borås, Sweden
Andrea Murari   Consorzio RFX-Associazione EURATOM ENEA
                                per la Fusione, Italy
Fionn Murtagh   University of Derby and Goldsmiths University
                                of London, UK
Ulf Norinder   Swedish Toxicology Sciences Research Center,
                                Sweden
Ilia Nouretdinov   Department of Computer Science, Royal Holloway,
                                University of London, Egham, Surrey, UK
Augusto Pereira   Laboratorio Nacional de Fusion, CIEMAT,
                                Madrid, Spain
Giuseppe Rattá   Laboratorio Nacional de Fusion, CIEMAT,
                                Madrid, Spain
Matilde Santos   Universidad Complutense de Madrid, Madrid, Spain
Victor Solovyev   Softberry Inc., USA
Rosanna Verde   Second University of Naples, Italy

Contents

Machine Learning

Invited Paper

Learning with Intelligent Teacher

Vladimir Vapnik[1,2] and Rauf Izmailov[3(✉)]

[1] Columbia University, New York, NY, USA
vladimir.vapnik@gmail.com
[2] AI Research Lab, Facebook, New York, NY, USA
[3] Applied Communication Sciences, Basking Ridge, NJ, USA
rizmailov@appcomsci.com

Abstract. The paper considers several topics on learning with privileged information: (1) general machine learning models, where privileged information is positioned as the main mechanism to improve their convergence properties, (2) existing and novel approaches to leverage that privileged information, (3) algorithmic realization of one of these (namely, knowledge transfer) approaches, and its performance characteristics, illustrated on simple synthetic examples.

Keywords: Intelligent teacher · Privileged information · Similarity control · Knowledge transfer · Knowledge representation · Frames · Support vector machines · SVM+ · Classification · Learning theory · Kernel functions · Regression

1 Introduction

The classical machine learning paradigm considers a simple scheme: given a set of training examples, find, in a given set of functions, the one that approximates the unknown decision rule in the best possible way. In such a paradigm, Teacher does not play an important role.

In human learning, however, the role of Teacher is important: along with examples, Teacher provides students with explanations, comments, comparisons, metaphors, and so on.

This paper considers the model of learning that includes the so-called Intelligent Teacher, who supplies Student with intelligent (privileged) information during training session. This privileged information exists for almost any learning problem and this information can significantly accelerate the learning process. In the learning paradigm called *Learning Using Privileged Information (LUPI)*, Intelligent Teacher provides additional (privileged) information x^* about training example x at the training stage (when Teacher interacts with Student). The

V. Vapnik—This material is based upon work partially supported by AFRL and DARPA under contract FA8750-14-C-0008 and the work partially supported by AFRL under contract FA9550-15-1-0502. Any opinions, findings and/or conclusions in this material are those of the authors and do not necessarily reflect the views of AFRL and DARPA.

© Springer International Publishing Switzerland 2016
A. Gammerman et al. (Eds.): COPA 2016, LNAI 9653, pp. 3–19, 2016.
DOI: 10.1007/978-3-319-33395-3_1

important point in this paradigm is that privileged information is *not* available at the test stage (when Student operates without supervision of Teacher). LUPI was initially introduced in [15,16]; subsequent work targeted various implementation issues of this paradigm [9] and its applications to a wide range of problems [3,4,10,12,19].

Formally, the classical paradigm of machine learning is described as follows: given a set of iid pairs (training data)

$$(x_1, y_1), \ldots, (x_\ell, y_\ell), \quad x_i \in X, \quad y_i \in \{-1, +1\}, \tag{1}$$

generated according to a fixed but unknown probability measure $P(x, y) = P(y|x)P(x)()$, find, in a given set of indicator functions $f(x, \alpha), \alpha \in \Lambda$, the function $y = f(x, \alpha_*)$ that minimizes the probability of incorrect classifications (incorrect values of $y \in \{-1, +1\}$). In this model, each vector $x_i \in X$ is a description of an example generated according to an unknown generator $P(x)$ of random vectors x_i, and $y_i \in \{-1, +1\}$ is its classification defined by Teacher according to an unknown conditional probability $P(y|x)$. The goal is to find the function $y = f(x, \alpha_*)$ that guarantees the smallest probability of incorrect classifications. That is, the goal is to find the function which minimizes the risk functional

$$R(\alpha) = \frac{1}{2} \int |y - f(x, \alpha)| dP(x, y), \tag{2}$$

in the given set of indicator functions $f(x, \alpha)$, $\alpha \in \Lambda$ when the probability measure $P(x, y) = P(y|x)P(x)$ is unknown but training data (1) are given.

The LUPI paradigm describes a more complex model: given a set of iid triplets

$$(x_1, x_1^*, y_1), \ldots, (x_\ell, x_\ell^*, y_\ell), \quad x_i \in X, \quad x_i^* \in X^*, \quad y_i \in \{-1, +1\}, \tag{3}$$

generated according to a fixed but unknown probability measure $P(x, x^*, y) = P(x*, y|x)P(x)$, find, in a given set of indicator functions $f(x, \alpha), \alpha \in \Lambda$, the function $y = f(x, \alpha_*)$ that guarantees the smallest probability of incorrect classifications (2). In this model, each vector $x_i \in X$ is a description of an example generated according to an unknown generator $P(x)$ of random vectors x_i, and Intelligent Teacher generates both its label $y_i \in \{-1, +1\}$ and the privileged information x_i^* using some unknown conditional probability function $P(x_i^*, y_i|x_i)$.

In the LUPI paradigm, we have exactly the same goal of minimizing (2) as in the classical paradigm, i.e., to find the best classification function in the admissible set. However, during the training stage, we have more information, i.e., we have triplets (x, x^*, y) instead of pairs (x, y) as in the classical paradigm. The additional information $x^* \in X^*$ belongs to space X^* which is, generally speaking, different from X.

The paper is organized in the following way. In Sect. 2, we outline general models of information theory and their relation to models of learning. In Sect. 3, we explain how privileged information can significantly accelerate the rate of learning (i.e., accelerate the convergence) when the notion of classical learning model is expanded appropriately to incorporate privileged information. In Sect. 4, we argue that structures in the space of privileged information reflect more fundamental properties of learning and thus can potentially improve the performance of learning methods even further. We outline a general *knowledge transfer* framework for realization of that improvement in Sect. 5. In Sect. 6, we present some specific algorithms implementing elements of that framework and illustrate their various properties on synthetic examples We conclude with Sect. 7, in which we summarize our results and outline potential next steps in this research.

2   Brute Force and Intelligent Models

In this section, we show how the general setting of machine learning problems creates a background for introduction of the concept of privileged information.

According to Kolmogorov [7], there exist three categories of integer numbers.

1. **Ordinary numbers:** those numbers n that we use in our everyday life. For simplicity, let these numbers be between 1 and one billion.
2. **Large numbers:** those numbers N that are between one billion and 2^n (where n belongs to the category of ordinary numbers).
3. **Huge numbers:** those numbers H that are greater than $2^N = 2^{2^n}$ (where N belongs to the category of large numbers).

Kolmogorov argued that the ordinary integers n correspond to the number of items we can handle realistically, say the number of examples in a learning problem. We cannot realistically handle large numbers (say large number of examples in a learning problem), but we can still treat them efficiently in our theoretical reasoning using mathematics; however, huge numbers are beyond our reach. In this paper, we describe methods that potentially might operate in huge sets of functions. In contrast to methods based on mathematical models and suitable for large numbers (which we call "brute force" methods), these methods include intelligent agents and thus can be viewed as "intelligent methods".

Basic Shannon Model. Suppose that our goal is to find one function among large number N of different functions by making ordinary number of queries that return the reply "yes" or "no" (thus providing one bit of information). Theoretically, we can find the desired function among N functions by making n queries, where $n = \log_2 N$ (for simplicity, we assume that N is an integer power of 2). Indeed, we can split the set of N functions into two subsets and make query to which subset the desired function belongs: to the first one (reply $+1$) or to the second one (reply -1). After obtaining the reply from the query, we can remove the subset which does not contain the desired function, split the remaining part into two subsets, and continue in the same fashion, removing half of the remaining functions after each reply. So after $n = \log_2 N$ queries we will find the function. It is easy to see that one cannot guarantee that it is possible to find the desired function by making less than

$$n = \log_2 N = \frac{\ln N}{\ln 2} \tag{4}$$

queries. This also means that one cannot find one function from the set of huge number $H = 2^N$ of functions: this would require to make too many (namely N) queries, which is unrealistic.

Basic Model using Language of Learning Theory. Let us repeat this reasoning for pattern recognition model. Suppose that our set $y = f(x, \alpha_t)$, $t = 1, \ldots, N$ is a finite set of binary functions in $x \in R^n$. That is, $f(x, \alpha_t) \in \{-1, +1\}$. Suppose that we can construct such vector $x_1 \in R^n$ that half of functions take value $f(x_1, \alpha_{t_1}^*) = +1$ and another half take value $f(x_1, \alpha_{t_1}) = -1$. Then the query for the label of vector x_1 provides the first element of training data (x_1, y_1). As before, we remove half of the functions that replied $-y_1$ and continue this process. After collecting at most $n = \ln N / \ln 2$ elements of training examples, we obtain the desired function.

First Modification of the Learning Model. To find the function in framework of basic model requires solution of a difficult problem: on any step of the procedure to find a vectors x_i that splits the remaining set of functions into two equal parts (suppose that such a vector exists). To simplify our model, consider the situation where vectors x are results of random iid trial with a fixed (but unknown) probability measure $p(x)$, and for any x we can query for its label y. After each query, we remove the functions that return $-y$ on x. The main problem for this model is to determine how many queries about labels one has to make[1] to find the function that is ε-close to the desired one with probability $1 - \eta$ (recall that the desired function is any function among those that do not make errors, and ε-closeness is defined with respect to measure $p(x)$). The answer to this problem constitutes a special case of the VC theory [13,14]: the number of the required queries is at most

$$\ell = \frac{\ln N - \ln \eta}{\varepsilon}. \tag{5}$$

This expression differs from bound (4) by a constant: $(\varepsilon)^{-1}$ instead of $(\ln 2)^{-1}$. After this number of queries, any function in the remaining set is ε-close to the desired one. This bound cannot be improved.

Second Modification of the Basic Model. So far, we considered the situation when the set of N functions includes the one that does not makes errors.

---

[1] In other words, how large should be the number ℓ of training examples $(x_1, y_1), \ldots, (x_\ell, y_\ell)$.

Now we relax this assumption: any function in our set of N functions can make errors. Our problem is to find the function than provides the smallest probability of error with respect to probability measure $p(x)$.

Now we cannot use the method for choosing the desired function defined in the first model: removing from the consideration the functions from the set that disagree with classification of query. We will use another (a more general) algorithm which selects such function among N of them that make the smallest number of disagreements with the query reply (i.e., minimizes the empirical loss) on the training set

$$(x_1, y_1), \ldots, (x_\ell, y_\ell).$$

In order to guarantee that we will select an ε-close function to the best in the set of N elements with probability $1 - \eta$, one has to make at most

$$\ell = \frac{\ln N - \ln \eta}{\varepsilon^2}$$

queries. Again, in this modification, the main term $\ln N$ remains the same but constant $(\varepsilon)^{-2}$ is different from the constant in (5). This bound cannot be improved.

Third Modification (VC Model). Consider now the set of functions $f(x, \alpha)$, $\alpha \in \Lambda$ with infinite number of elements. Generally speaking, in this situation one cannot guarantee that it is possible to obtain a good approximation even if we have a large number of training examples. Recall that in the more simple situation with a set that contains finite but huge number of functions $H = 2^{2^n}$, one needs 2^n examples, which is far beyond our reach. Nevertheless, if infinite set of functions has finite VC dimension $VCdim$, then ε-close solution can be found with probability $1 - \eta$ using at most

$$\ell = \frac{VCdim - \ln \eta}{\varepsilon}$$

observations, if the desired function does not make errors; otherwise, if errors are allowed,

$$\ell = \frac{VCdim - \ln \eta}{\varepsilon^2}$$

observations are required. Note that this bound matches the form of bound (5), where the value of VC dimension replaces the logarithm of the number of functions in the set. This bound cannot be improved.

The finiteness of the VC dimension of the set of functions defines the necessary and sufficient conditions of learnability (consistency) of empirical risk minimization method. This means that VC dimension characterizes not just the *quantity* of elements of the set, it characterizes something else, namely, the *measure of diversity* of the set of functions: the set of functions must be not too diverse.

The structural risk minimization principle that uses structure on the nested subsets of functions with finite VC dimension (defined on the sets of functions which closure can have infinite VC dimension) guarantees convergence of risk to the best possible risk for this structure [13,14].

To summarize, we have outlined the best bounds for general machine learning models and stated that they cannot be improved. In other words, in order to improve these bounds, the models themselves will have to be changed. The specific model change that we are concerned with in this paper is provided by the notion of privileged information, which is described and explored in the subsequent sections.

3   Privileged Information As Learning Acceleration

The learning models described in the previous section can be solved by different methods. In particular, SVM algorithms with universal kernels realize structural risk minimization method and thus are universally consistent. This means that the VC theory completely solves the problem of learning from examples providing not only the necessary and sufficient conditions of learnability but also an effective practical algorithm for machine learning. The rate described by this theory cannot be improved essentially (without additional information).

The intriguing question in VC theory was why the number of examples one needs to construct ε-close hyperplane in separable case (when training data can be separated without errors) and unseparable case (when training data cannot be separated without errors) vary so much in their corresponding constants (ε^{-1} and ε^{-2}).

For SVM algorithm, this effect can be explained by noticing that, in the separable case, using ℓ examples one has to estimate n parameters w of hyperplane, while in the non-separable case, one has to estimate, along with parameters n of hyperplane w, the additional ℓ values of slacks (making the total number of parameters to be estimated larger than number of examples). This, however, can addressed by a special SVM+ algorithm within the LUPI framework [15,16]. In that framework, Intelligent Teacher supplies Student with triplets

$$(x_1, x^*, y_1), \ldots, (x_\ell, x_\ell^*, y_\ell)$$

where $x_i \in X^*$, whereas, in the classical setting of the problem, Student uses training pairs

$$(x_1, y_1), \ldots, (x_\ell, y_\ell)$$

where vector $x_i \in X$ is generated by the generator of random events $p(x)$ and Teacher supplies Student with the label $y_i \in \{-1, +1\}$. In contrast to classical setting, in the LUPI paradigm, Intelligent Teacher supplies Student with triplets (x_i, x_i^*, y_i) where vector $x_i^* \in X^*$ and label y are generated by conditional probability $p(x^*, y|x)$. Formally, by providing both vector $x^* \in X^*$ and label y_i for any example x_i, Intelligent Teacher can supply Student with *more than one bit of information*, so the rate of convergence can be faster.

Indeed, as was shown in [15,16], this SVM+ approach in LUPI can improve the constant from ε^{-2} to ε^{-1}. The recent LUPI papers [17,18] introduced more important approaches that could be potentially used for further improvement of convergence. In order to use such mechanisms effectively, Intelligent Teacher has to possess some knowledge that can describe physical model of events better than x. In the subsequent sections, we describe these ideas in greater detail.

4   Space of Privileged Information

Let us suppose that Intelligent Teacher has some knowledge about the solution of a specific pattern recognition problem and would like to transfer this knowledge to Student. For example, Teacher can reliably recognize cancer in biopsy images (in a pixel space X) and would like to transfer this skill to Student.

Formally, this means that Teacher has some function $y = f_0(x)$ that distinguishes cancer ($f_0(x) = +1$ for cancer and $f_0(x) = -1$ for non-cancer) in the pixel space X. Unfortunately, Teacher does not know this function explicitly (it only exists as a neural net in Teacher's brain), so how can Teacher transfer this construction to Student? Below, we describe a possible mechanism for solving this problem; we call this mechanism *knowledge transfer*.

Suppose that Teacher believes in some theoretical model on which the knowledge of Teacher is based. For cancer model, he or she believes that it is a result of uncontrolled multiplication of the cancer cells (cells of type B) which replace normal cells (cells of type A). Looking at a biopsy image, Teacher tries to generate privileged information that reflects his or her belief in development of such s process; Teacher can describe the image as:

Aggressive proliferation of cells of type B into cells of type A.

If there are no signs of cancer activity, Teacher may use the description

Absence of any dynamics in the of standard picture.

In uncertain cases, Teacher may write

There exist small clusters of abnormal cells of unclear origin.

In other words, Teacher is developing a specialized language that is appropriate for description x_i^* of cancer development using the model he believes in. Using this language, Teacher supplies Student with privileged information x_i^* for the image x_i by generating training triplets

$$(x_1, x_1^*, y_1), \ldots, (x_\ell, x_\ell^*, y_\ell). \tag{6}$$

The first two elements of these triplets are descriptions of an image in two languages: in language X (vectors x_i in pixel space), and in language X^* (vectors x_i^* in the space of privileged information), developed for Teacher's understanding of cancer model.

Note that the language of pixel space is universal (it can be used for description of many different visual objects; for example, in the pixel space, one can distinguish between male and female faces), while the language used for describing privileged information is very specialized: it reflects just a model of cancer development. This has an important consequence: the set of admissible functions in space X has to be rich (has a large VC dimension), while the set of admissible functions in space X^* may be not rich (has a small VC dimension).

One can consider two related pattern recognition problems using triplets (6):

1. The problem of constructing a rule $y = f(x)$ for classification of biopsy in the pixel space X using data

$$(x_1, y_1), \ldots, (x_\ell, y_\ell). \tag{7}$$

2. The problem of constructing a rule $y = f^*(x^*)$ for classification of biopsy in the space X^* using data

$$(x_1^*, y_1), \ldots, (x_\ell^*, y_\ell). \tag{8}$$

Suppose that language X^* is so good that it allows to create a rule $y = f_\ell^*(x^*)$ that classifies vectors x^* corresponding to vectors x with the same level of accuracy as the best rule $y = f_\ell(x)$ for classifying data in the pixel space.[2]

Since the VC dimension of the admissible rules in a special space X^* is much smaller than the VC dimension of the admissible rules in the universal space X and since, the number of examples ℓ is the same in both cases, the bounds on error rate for the rule $y = f_\ell^*(x^*)$ in X^* will be better[3] than those for the rule $y = f_\ell(x)$ in X. That is, generally speaking, the classification rule $y = f_\ell^*(x^*)$ will be more accurate than classification rule $y = f_\ell(x)$.

As a result, the following problem arises: how one can use the knowledge of the rule $y = f_\ell^*(x^*)$ in space X^* to improve the accuracy of the desired rule $y = f_\ell(x)$ in space X? A general framework for that is outlined in the next section.

5   Knowledge Transfer from Privileged Space

As already described, knowledge transfer approach deals with iid training examples generated by some unknown generator $P(x), x \in X$ and Intelligent Teacher who supplies vectors x with information $(x^*, y|x)$ according to some (unknown) *Intelligent generator* $P(x^*, y|x), x^* \in X^*, y \in \{-1, +1\}$, forming training triplets

$$(x_1, x_1^*, y_1), \ldots, (x_\ell, x_\ell^*, y_\ell). \tag{9}$$

---

[2] The rule constructed in space X^* cannot be better than the best possible rule in space X, since all information originates in space X.

[3] According to VC theory, the guaranteed bound on accuracy of the chosen rule depends only on two factors: frequency of errors on training set and VC dimension of admissible set of functions.

Consider two pattern recognition problems in decision and privileged spaces:

1. *Pattern recognition problem defined in space X*: Using data, $(x_1, y_1), \ldots,$ (x_ℓ, y_ℓ), find in set of functions $f(x, \alpha), \alpha \in \Lambda$ the rule $y = \text{sgn}\{f_\ell(x)\}$ that minimizes the probability of test errors (in space X).
2. *Pattern recognition problem defined in space X^**: Using data, $(x_1^*, y_1), \ldots,$ (x_ℓ^*, y_ℓ), find in set of functions $f^*(x^*, \alpha^*), \alpha^* \in \Lambda^*$ the rule $y = \text{sgn}\{f_\ell^*(x^*)\}$ that minimizes the probability of test errors (in space X^*).

Suppose that, in space X^*, one can find a rule $y = \text{sgn}\{f_\ell^*(x^*)\}$ that is, with probability $1 - \eta$, is better than the corresponding rule $y = \text{sgn}\{f_\ell(x)\}$ in space X. Also, suppose that we are looking for our rule in the form

$$f_\ell^*(x^*) = \sum_{i=1}^{\ell} y_i \alpha_i K_i^*(x^*) + b^*, \tag{10}$$

where α_i^*, $i = 1, \ldots, \ell$ and b^* are parameters, and K_i are some functions in X^*. The question is whether the knowledge of a good rule (10) in space X^* can be used to find a good rule

$$s = f_\ell(x) = \sum_{i=1}^{\ell} y_i \alpha_i K(x_i, x) + b \tag{11}$$

in space X.

As was described in the previous section for the problem of cancer diagnostics, since pixel space X is universal and space of descriptions X^* reflects just the model of cancer development[4], the VC dimension of admissible set of functions in X space has to be much bigger than VC dimension of admissible set of functions in X^*. Therefore, with probability $1 - \eta$, the guaranteed quality of the rule constructed from ℓ examples in space X^* will be better than the quality of the rule constructed from ℓ examples in space X. That is why a transfer of a rule from space X^* into space X can be helpful.

In order to describe knowledge transfer, consider two fundamental concepts of knowledge representation used in Artificial Intelligence [1]:

1. Frames (fragments) of the knowledge.
2. Structural connections of the frames (fragments) in the knowledge.

The actual realization of frames and structures of knowledge can be done using different methods. For example, we can call the *frames in the knowledge* the smallest number of the vectors $u_1^* \ldots, u_m^*$ from space X^* that can approximate[5] the main part of the rule (10):

$$f_\ell^*(x^*) - b^* = \sum_{i=1}^{\ell} y_i \alpha_i^* K^*(x_i^*, x^*) \approx \sum_{k=1}^{m} \beta_k^* K^*(u_k^*, x^*). \tag{12}$$

---

[4] In this example generator $P(x^*, y|x)$ is intelligent since for any *picture* of the event x it describes the *essence* of the event. Using description of the essence of the event makes classification of the event an easy problem.

[5] In machine learning, they are called the reduced number of support vectors [2].

We then call the functions $K^*(u_k^*, x^*)$, $k = 1, \ldots, m$ the *frames* (fragments) of knowledge. Our knowledge

$$f_\ell^*(x^*) = \sum_{k=1}^m \beta_k^* K^*(u_k^*, x^*) + b$$

is defined as a linear combination of the frames.

In the described terms, knowledge transfer from X^* into X requires the following:

1. To find the fundamental elements of knowledge $u_1^*, \ldots, u_m^*$ in space X^*.
2. To find frames (m functions) $K^*(u_1^*, x^*), \ldots, K^*(u_m^*, x^*)$ in space X^*.
3. To find the functions $\phi_1(x), \ldots, \phi_m(x)$ in space X such that

$$\phi_k(x_i) \approx K^*(u_k^*, x_i^*) \tag{13}$$

holds true for almost all pairs (x_i, x_i^*) generated by Intelligent Teacher that uses some (unknown) generator $P(x^*, y|x)$.

Note that the capacity of the set of functions from which $\phi_k(x)$ are to be chosen can be smaller than that of the capacity of the set of functions from which the classification function $y = f_\ell(x)$ is chosen (function $\phi_k(x)$ approximates just one fragment of knowledge, not the entire knowledge as function $y = f_\ell^*(x^*)$, which is a linear combination (12) of frames). Also, estimates of all the functions $\phi_1(x), \ldots, \phi_m(x)$ are done using different pairs as training sets of the same size ℓ. We hope that transfer of m fragments of knowledge from space X^* into space X can be done with higher accuracy than estimating function $y = f_\ell(x)$ from data (7).

After finding approximation of frames in space X, the knowledge about the rule obtained in space X^* can be approximated in space X as

$$f_\ell(x) \approx \sum_{k=1}^m \delta_k \phi_k(x) + b^*,$$

where coefficients $\delta_k = \alpha_k^*$ (taken from (10)) if approximations (13) are accurate. Otherwise, coefficients δ_k can be estimated from the training data.

More generally, in order to transfer knowledge from space X^* to space X one has to make the following two transformations in the training triplets (9):

1. Transform n-dimensional vectors of $x_i = (x_i^1, \ldots, x_i^n)^T$ into k-dimensional vectors $\mathcal{F}x_i = (\phi_1(x_i), \ldots, \phi_k(x_i))^T$. In order to transform vector x, one constructs m-dimensional space as follows: for any frame $K^*(x^*, x_s^*), s = 1, \ldots, k$ in space X^*, one construct it image (function) $\phi_s(x)$ in space X that defines the relationship

$$\phi_s(x) = \int K(x_s^*, x^*) P(x^*|x) dx^*, \qquad s = 1, \ldots, k.$$

This requires to solve the following regression estimation problem: given data

$$(x_1, z_1^s), \ldots, (x_\ell, z_\ell^s), \quad \text{where} \quad z_i^s = K(x_s^*, x_i^*),$$

find regression functions $\phi_s(x)$, $s = 1, \ldots, k$, forming the space
$\mathcal{F}(x) = (\phi_1(x), \ldots, \phi_k(x))^T$.

2. Use the target values s_i^* obtained for x_i^* in rule (10) instead of the values y_i given for x_i in triplet (9), i.e., replace target value y_i in triplets (9) with scores s_i^* given (10).

Thus the knowledge transfer algorithm transforms the training triplet[6]

$$((\mathcal{F}x_1, x_1^*, s_1^*), \ldots, (\mathcal{F}x_\ell, x_\ell^*, s_\ell^*)), \tag{14}$$

and then uses triplets (14) instead of triplets (9).

6 Feature-Based Algorithm for Knowledge Transfer

In this section, we present scalable algorithms of knowledge transfer in LUPI based on multivariate regressions of privileged features as functions of decision variables; we also illustrate the algorithms' performance and their properties on synthetic examples.

We assume again that we are given a set of iid triplets

$$(x_1, x_1^*, y_1), \ldots, (x_\ell, x_\ell^*, y_\ell), \quad x_i \in X = R^n, \quad x_i^* \in X^* = R^m, \quad y_i \in \{-1, +1\},$$

generated according to a fixed but unknown probability measure $P(x, x^*, y)$. Our training dataset consists of ℓ decision vectors $x_1, \ldots, x_\ell$ from n-dimensional decision space $X = R^n$ and corresponding ℓ privileged vectors $x_1^*, \ldots, x_\ell^*$ from m-dimensional privileged space $X^* = R^m$.

In order to create knowledge transfer from space X^*, we use training data $x_1, \ldots, x_\ell$ to construct m multivariate regression functions $\phi_i(x^1, \ldots, x^n)$, where $i = 1, \ldots, m$, from n-dimensional decision space X to each of our m privileged features. Various types of regression could be used for that purpose, such as linear ridge regression or nonlinear kernel regression. After those regressions ϕ_i are constructed, we replace, for each $j = 1, \ldots, \ell$ and each $i = 1, \ldots, m$, the ith coordinate of jth privileged vector x_j^* with its regressed approximation $\phi_i(x_j^1, \ldots, x_j^n)$. In the next step, we construct the modified training dataset, consisting of m-dimensional regression-based replacements of privileged vectors. As a result, our modified training data will form the matrix

$$\begin{pmatrix} y_1 & \phi_1(x_1^1, \ldots, x_1^n) & \cdots & \phi_m(x_1^1, \ldots, x_1^n) \\ y_2 & \phi_1(x_2^1, \ldots, x_2^n) & \cdots & \phi_m(x_2^1, \ldots, x_2^n) \\ \cdots & \cdots & \cdots & \cdots \\ y_\ell & \phi_1(x_\ell^1, \ldots, x_\ell^n) & \cdots & \phi_m(x_\ell^1, \ldots, x_\ell^n) \end{pmatrix}.$$

Then, we apply some standard SVM algorithm to this modified training data and construct an m-dimensional decision rule. This rule can be used to classify any n-dimensional test vector $z = (z^1, \ldots, z^n)$ by executing the following steps:

---

[6] In the simplified version, pairs $(\mathcal{F}x_i, s_i^*)$, $i = 1, \ldots, \ell$..

1. Using previously constructed (at the training stage) m multivariate regressions $\phi_1, \ldots, \phi_m$, compute m approximations to the unavailable privileged variables (coordinates) and form the m-dimensional vector

$$z^* = (\phi_1(z^1, \ldots, z^n), \phi_2(z^1, \ldots, z^n), \ldots, \phi_m(z^1, \ldots, z^n)).$$

2. Apply the constructed m-dimensional SVM decision rule to this m-dimensional augmented test vector z^*.

The described algorithm of knowledge transfer completely solves the main scalability problem of SVM+ algorithm, which was not practical for problems with more than several hundred training samples. Indeed, for larger number of samples, the SVM+ matrix for quadratic programming becomes ill-conditioned and larger number of parameters makes the problem of SVM+ parameter selection very time consuming [9]. In contrast to that, while the described knowledge transfer algorithm requires an additional step of calculating m multivariate regressions, which takes some limited time, this regression computation is performed only once during the whole process of parameter optimization (i.e., during grid search), and, most importantly, the augmented training data are then processed with any standard scalable SVM implementation.

In order to illustrate properties of the described knowledge transfer LUPI algorithm, consider its performance on the following simple synthetic example.

For training dataset, we generated ℓ two-dimensional random points (x^1, x^2), uniformly distributed in the square $[-1, +1] \times [-1, +1]$. Each point (x^1, x^2) was labeled with $y = \text{sgn}(x^1 + x^2)$. Both dimensions of these points were treated as standard decision features. In addition, for each point (x^1, x^2), we generated the value $x^3 = x^1 + x^2 + \varepsilon W$, where ε is the noise parameter, and W is an $N(0, 1)$-distributed random number; x^3 was treated as a privileged variable. Therefore, in this model, the privileged variable x^3 is more or less closely (depending on the noise level ε) related to the label of the decision vector (x^1, x^2).

We considered the following three types of classification scenarios:

- **SVM on decision features:** Training points (x^1, x^2) belong to the two-dimensional decision space, and RBF SVM is used to create the decision rule.
- **Knowledge transfer LUPI:** Training points (x^1, x^2) belong to the two-dimensional decision space, while privileged feature $(x_1^3, \ldots, x_\ell^3)^T$ belongs to the one-dimensional privileged space; knowledge transfer from privileged feature x^3 to the space of decision features $(x_1^1, \ldots, x_\ell^1)^T$ and $(x_1^2, \ldots, x_\ell^2)^T$ is realized with linear ridge regression. After augmenting x^1 and x^2 with regressed value of x^3, we construct the RBF SVM decision rule in the one-dimensional decision space.
- **SVM on privileged features:** Training points (x^3) belong to the one-dimensional decision space, and RBF SVM is used to create the decision rule.

For each of these scenarios, the error rate of the constructed decision rule was measured on the test dataset, generated according to the same distribution and containing (for statistical reliability of results) 10,000 two-dimensional points (x^1, x^2). In our experiments, for each value of ℓ (selected as $10, 20, 40$) and each value of ε (selected as $0.01, 0.1, 1.0$), we generated 10 random realizations of training datasets of ℓ samples each. For each of these $10 \times 3 \times 3 = 90$ datasets, we ran all three classification scenarios (SVM on decision features, Knowledge trasnfer LUPI, and SVM on privileged features). Two parameters for RBF kernels (utilized in all three scenarios), namely SVM penalty parameter C and RBF kernel parameter γ, were selected using 6-fold cross-validation error rate over the two-dimensional grid of both parameters C and γ. In that grid, $\log_2(C)$ ranged of from -5 to $+5$ with step 0.5, and $\log_2(\gamma)$ ranged $+6$ to -6 with step 0.5 (thus the whole grid consisted of $21 \times 25 = 525$ pairs of tested parameters C and γ).

Table 1. Performance of SVMs and LUPI on synthetic example.

noise=0.01			
	training size 10	training size 20	training size 40
SVM on decision features	22.53 %	7.12 %	5.45 %
Knowledge transfer LUPI	10.10 %	2.32 %	1.77 %
SVM on privileged features	10.07 %	2.32 %	1.94 %
noise=0.1			
	training size 10	training size 20	training size 40
SVM on decision features	22.53 %	7.12 %	5.45 %
Knowledge transfer LUPI	10.22 %	2.30 %	2.06 %
SVM on privileged features	9.97 %	2.72 %	2.07 %
noise=1.0			
	training size 10	training size 20	training size 40
SVM on decision features	22.53 %	7.12 %	5.45 %
Knowledge transfer LUPI	18.24 %	5.74 %	3.44 %
SVM on privileged features	22.80 %	15.97 %	13.23 %

The averaged (over 10 realizations) error rates are shown in Table 1. The collected results suggest the following conclusions:

1. Knowledge Transfer LUPI improves the performance of Standard SVM on decision features (often significantly, in relative terms) in all of the considered scenarios. This relative improvement depends on interplay of noise and size of training sample.
2. For larger values of noise and/or larger training sizes, we observe that Knowledge Transfer LUPI can be even *better* that SVM on privileged features. While appearing counter-intuitive (an approximated (regressed) value turns out to be better for classification that the real one), this effect is due to the nature of synthetic distribution we used for this example. Indeed, for a large noise, the regressed privileged variable approximates the label function $\operatorname{sgn}(x^1 + x^2)$

much more accurately (especially for large training size) than the actual data available during the training (since the accurate regression filters out most of the noise in the data). It also demonstrates the value of proper learning the structures of privileged space (with linear regression, in this example): if we learn these structures well, we might be able to improve performance significantly, even beyond the one delivered by SVM on privileged features.

Note that this is just one possible way to apply the idea of feature-based knowledge transfer. In many realistic examples, it is prudent not to switch completely from decision features to regressed privileged ones, but rather use both types of features in concatenation, thus forming the matrix of augmented training data

$$\begin{pmatrix} y_1 & x_1^1 & \cdots & x_1^n & \phi_1(x_1^1, \ldots, x_1^n) & \cdots & \phi_m(x_1^1, \ldots, x_1^n) \\ y_2 & x_2^1 & \cdots & x_2^n & \phi_1(x_2^1, \ldots, x_2^n) & \cdots & \phi_m(x_2^1, \ldots, x_2^n) \\ \cdots\cdots\cdots\cdots & & & \cdots & & \cdots & \cdots \\ y_\ell & x_\ell^1 & \cdots & x_\ell^n & \phi_1(x_\ell^1, \ldots, x_\ell^n) & \cdots & \phi_m(x_\ell^1, \ldots, x_\ell^n) \end{pmatrix}.$$

In this version of knowledge transfer LUPI, we apply some standard SVM algorithm to this augmented training data and construct an $(n + m)$-dimensional decision rule. This rule is then used to classify any test n-dimensional test vector $z = (z^1, \ldots, z^n)$ by executing the following steps:

1. Using previously constructed (at the training stage) m multivariate regressions $\phi_1, \ldots, \phi_m$, compute m approximations to the unavailable privileged variables (coordinates) and form the m-dimensional vector

$$z^* = (\phi_1(z^1, \ldots, z^n), \phi_2(z^1, \ldots, z^n), \ldots, \phi_m(z^1, \ldots, z^n)).$$

2. Concatenate the n-dimensional test vector z with this m-dimensional vector z^* to form augmented $(n + m)$-dimensional vector

$$(zz^*) = (z^1, \ldots, z^n, \phi_1(z^1, \ldots, z^n), \phi_2(z^1, \ldots, z^n), \ldots, \phi_m(z^1, \ldots, z^n))$$

3. Apply the $(n+m)$-dimensional SVM decision rule to this $(n+m)$-dimensional augmented test vector (zz^*).

In order to illustrate this version of knowledge transfer LUPI, we explored another synthetic dataset, derived from dataset "Parkinsons" in [8]. Since none of 22 features of "Parkinsons" dataset is privileged, we created several artificial scenarios emulating the presence of privileged information in that dataset. Specifically, we ordered "Parkinsons" features according to the values of their mutual information (with first features having the lowest mutual information, while the last features having the largest one). Then, for several values of parameter k, we treated the last k features as privileged ones, with the first $22 - k$ features being treated as decision ones. Since our ordering was based on mutual information, these experiments corresponded to privileged spaces of various dimensions and various relevance levels for classification. For each considered value of k, we generated 20 pairs of training and test subsets, containing, respectively 75 % and 25 % of elements of the "Parkinsons" dataset. For each of these pairs, we considered the following four types of classification scenarios:

- RBF SVM on $22 - k$ decision features;
- Knowledge transfer LUPI based on constructing k multivariate regressions from $22 - k$ decision features to each of k privileged ones, replacing the corresponding values in privileged vectors with their regressed approximations, and training RBF SVM on the augmented dataset consisting of 22 features;
- RBF SVM on k privileged features;
- RBF SVM on k all features.

In all these experiments, the parameters for RBF kernels were selected in the same way as for previous synthetic example.

Table 2. Performance of SVMs and LUPI on modified "Parkinsons" example.

k	SVM on decision features	Knowledge transfer LUPI	SVM on privileged features	SVM on all features
1	9.18 %	8.77 %	21.12 %	7.92 %
2	11.33 %	10.21 %	18.37 %	7.92 %
3	12.24 %	9.67 %	12.96 %	7.92 %
4	15.20 %	13.47 %	13.06 %	7.92 %
5	16.22 %	13.78 %	12.40 %	7.92 %
6	16.35 %	12.36 %	11.71 %	7.92 %
7	16.81 %	13.55 %	11.63 %	7.92 %
8	17.02 %	14.12 %	11.12 %	7.92 %
9	17.50 %	13.16 %	10.98 %	7.92 %
10	17.91 %	15.61 %	10.71 %	7.92 %

The averaged (over 20 realizations) error rates for these scenarios are shown in Table 2. The collected results suggest the following conclusions:

1. Knowledge Transfer LUPI improves the performance of Standard SVM on decision features (often significantly, in relative terms) in all of the considered scenarios. The error rates of LUPI are between SVMs constructed on decision features and on all features. In other words, if the error rate of SVM on decision features is B, while the error rate of SVM on all features is C, the error rate A of LUPI satisfies the bounds $C < A < B$. So one can evaluate the efficiency of LUPI approach by computing the metric $(B - A)/(B - C)$, which describes how much of the performance gap $B - C$ can be recovered by LUPI. In Table 2, this metric varies between 23 % and 59 %. Generally, in realistic examples, the typical value for this LUPI efficiency metric is in the ballpark of 35 %. Also note that if the gap $B - C$ is small compared to

C, it means that the privileged information is not particularly relevant; in that case, it is likely hopeless to apply LUPI anyway: there is little space for improvement for that. It is probably safe to start looking for LUPI solution if the gap $B - C$ is at least $1.5 - 2$ times larger than C.

2. The error rate of SVM on privileged features only becomes better than that of SVM on decision features for values of k larger than 3. This suggests that it is safer to rely on both decision and regressed privileged features in LUPI construction, since privileged features alone may not be sufficient to replace the classification information contained in decision features.

7   Conclusions

In this paper, we presented several properties of privileged information including its role in machine learning, its structure, and its applications. We extended the previous research in the area of privileged information by highlighting structures in the space of privileged information and various mechanisms that can leverage those structures for producing better solutions of pattern recognition problems. In particular, we presented a simple scalable algorithm for knowledge transfer, which avoids the scalability problem of current SVM+ implementations of LUPI. This algorithm is just a first step in the proposed direction, and its further improvements (especially concerning proper selection of relevant privileged features) will be the subject of future work.

References

1. Brachman, R., Levesque, H.: Knowledge Representation and Reasoning. Morgan Kaufmann, San Francisco (2004)
2. Burges, C.: Simplified support vector decision rules. In: 13th International Conference on Machine Learning, pp. 71–77 (1996)
3. Fouad, S., Tino, P., Raychaudhury, S., Schneider, P.: Incorporating privileged information through metric learning. IEEE Trans. Neural Netw. Learn. Syst. **24**, 1086–1098 (2013)
4. Ilin, R., Streltsov, S., Izmailov, R.: Learning with privileged information for improved target classification. Int. J. Monit. Surveill. Technol. Res. **2**(3), 5–66 (2014)
5. Izmailov, R., Vapnik, V., Vashist, A.: Multidimensional splines with infinite number of knots as SVM Kernels. In: International Joint Conference on Neural Networks, pp. 1096–1102. IEEE Press, New York (2013)
6. Kimeldorf, G., Wahba, G.: some results on Tchebycheffian spline functions. J. Math. Anal. Appl. **33**, 82–95 (1971)
7. Kolmogorov, A.: Mathematics as a Profession. Nauka, Moscow (1988). (in Russian)
8. Lichman, M.: UCI machine learning repository. University of California, School of Information and Computer Science, Irvine, CA (2013). http://archive.ics.uci.edu/ml
9. Pechyony, D., Izmailov, R., Vashist, A., Vapnik, V.: SMO-style algorithms for learning using privileged information. In: 2010 International Conference on Data Mining, pp. 235–241 (2010)

10. Ribeiro, B., Silva, C., Chen, N., Vieirac, A., das Nevesd, J.C.: Enhanced default risk models with SVM+. Expert Syst. Appl. **39**, 10140–10152 (2012)
11. Schölkopf, B., Herbrich, R., Smola, A.J.: A generalized representer theorem. In: Helmbold, D.P., Williamson, B. (eds.) COLT 2001 and EuroCOLT 2001. LNCS (LNAI), vol. 2111, pp. 416–426. Springer, Heidelberg (2001)
12. Sharmanska, V., Lampert, C.: Learning to rank using privileged information. In: 2013 IEEE International Conference on Computer Vision (ICCV), pp. 825–832 (2013)
13. Vapnik, V.: The Nature of Statistical Learning Theory. Springer-Verlag, New York (1995)
14. Vapnik, V.: Statistical Learning Theory. Wiley, New York (1998)
15. Vapnik, V.: Estimation of Dependencies Based on Empirical Data, 2nd edn. Springer, New York (2006)
16. Vapnik, V., Vashist, A.: A new learning paradigm: learning using privileged information. Neural Netw. **22**, 546–557 (2009)
17. Vapnik, V., Izmailov, R.: Learning with intelligent teacher: similarity control and knowledge transfer. In: Gammerman, A., Vovk, V., Papadopoulos, H. (eds.) SLDS 2015. LNCS, vol. 9047, pp. 3–32. Springer, Heidelberg (2015)
18. Vapnik, V., Izmailov, R.: Learning using privileged information: similarity control and knowledge transfer. J. Mach. Learn. Res. **16**, 2023–2049 (2015)
19. Yang, H., Patras, I.: Privileged information-based conditional regression forest for facial feature detection. In: 2013 IEEE International Conference on Automatic Face and Gesture Recognition, pp. 1–6 (2013)

Theory of Conformal Prediction

Criteria of Efficiency for Conformal Prediction

Vladimir Vovk[1]([✉]), Valentina Fedorova[2], Ilia Nouretdinov[1],
and Alexander Gammerman[1]

[1] Computer Learning Research Centre, Royal Holloway,
University of London, Egham, Surrey, UK
volodya.vovk@gmail.com, {ilia,alex}@cs.rhul.ac.uk
[2] Yandex, Moscow, Russia
alushaf@gmail.com

Abstract. We study optimal conformity measures for various criteria
of efficiency in an idealised setting. This leads to an important class of
criteria of efficiency that we call probabilistic; it turns out that the most
standard criteria of efficiency used in literature on conformal prediction
are not probabilistic.

Keywords: Conformal prediction · Predictive efficiency · Informational
efficiency

1   Introduction

Conformal prediction is a method of generating prediction sets that are guaranteed to have a prespecified coverage probability; in this sense conformal predictors have guaranteed validity. Different conformal predictors, however, widely
differ in their efficiency, by which we mean the narrowness, in some sense, of
their prediction sets. Empirical investigation of the efficiency of various conformal predictors is becoming a popular area of research: see, e.g., [1,11] (and the
COPA Proceedings, 2012–2015). This paper points out that the standard criteria
of efficiency used in literature have a serious disadvantage, and we define a class
of criteria of efficiency, called "probabilistic", that do not share this disadvantage. In two recent papers [3,5] two probabilistic criteria have been introduced,
and in this paper we introduce two more and argue that probabilistic criteria
should be used in place of more standard ones. We concentrate on the case of
classification only (the label space is finite).

Surprisingly few criteria of efficiency have been used in literature, and even
fewer have been studied theoretically. We can speak of the efficiency of individual
predictions or of the overall efficiency of predictions on a test sequence; the latter
is usually (in particular, in this paper) defined by averaging the efficiency over

---

A preliminary version of this paper was published as Working Paper 11 of the On-line
Compression Modelling project (New Series), http://alrw.net, in April 2014.

© Springer International Publishing Switzerland 2016
A. Gammerman et al. (Eds.): COPA 2016, LNAI 9653, pp. 23–39, 2016.
DOI: 10.1007/978-3-319-33395-3_2

the individual test examples, and so in this introductory section we only discuss the former. This section assumes that the reader knows the basic definitions of the theory of conformal prediction, but they will be given in Sect. 2, which can be consulted now.

The two criteria for efficiency of a prediction that have been used most often in literature (in, e.g., the references given above) are:

- The confidence and credibility of the prediction (see, e.g., [14], p. 96; introduced in [12]). This criterion does not depend on the choice of a significance level ϵ.
- Whether the prediction is a singleton (the ideal case), multiple (an inefficient prediction), or empty (a superefficient prediction) at a given significance level ϵ. This criterion was introduced in [10], Sect. 7.2, and used extensively in [14].

The other two criteria that have been used are the sum of the p-values for all potential labels (this does not depend on the significance level) and the size of the prediction set at a given significance level: see the papers [3,5].

In this paper we introduce six other criteria of efficiency (Sect. 2). We then discuss (in Sects. 3, 4 and 5) the conformity measures that optimise each of the ten criteria when the data-generating distribution is known; this sheds light on the kind of behaviour implicitly encouraged by the criteria even in the realistic case where the data-generating distribution is unknown. As we point out in Sect. 5, probabilistic criteria of efficiency are conceptually similar to "proper scoring rules" in probability forecasting [2,4], and this is our main motivation for their detailed study in this paper. After that we briefly illustrate the empirical behaviour of two of the criteria for standard conformal predictors and a benchmark data set (Sect. 6).

We only consider the case of randomised ("smoothed") conformal predictors: the case of deterministic predictors may lead to packing problems without an explicit solution (this is the case, e.g., for the N criterion defined below). The situation here is analogous to the Neyman–Pearson lemma: cf. [7], Sect. 3.2.

2   Criteria of Efficiency for Conformal Predictors and Transducers

Let $\mathbf{X}$ be a measurable space (the *object space*) and $\mathbf{Y}$ be a finite set equipped with the discrete σ-algebra (the *label space*); the *example space* is defined to be $\mathbf{Z} := \mathbf{X} \times \mathbf{Y}$. A *conformity measure* is a measurable function A that assigns to every finite sequence $(z_1, \ldots, z_n) \in \mathbf{Z}^*$ of examples a same-length sequence $(\alpha_1, \ldots, \alpha_n)$ of real numbers and that is equivariant with respect to permutations: for any n and any permutation π of $\{1, \ldots, n\}$,

$$(\alpha_1, \ldots, \alpha_n) = A(z_1, \ldots, z_n) \implies \left(\alpha_{\pi(1)}, \ldots, \alpha_{\pi(n)}\right) = A\left(z_{\pi(1)}, \ldots, z_{\pi(n)}\right).$$

The *conformal predictor* determined by A is defined by

$$\Gamma^\epsilon(z_1, \ldots, z_l, x) := \{y \mid p^y > \epsilon\}, \tag{1}$$

where $(z_1, \ldots, z_l) \in \mathbf{Z}^*$ is a training sequence, x is a test object, $\epsilon \in (0, 1)$ is a given *significance level*, for each $y \in \mathbf{Y}$ the corresponding *p-value* p^y is defined by

$$p^y := \frac{1}{l+1} \left| \{ i = 1, \ldots, l+1 \mid \alpha_i^y < \alpha_{l+1}^y \} \right|$$

$$+ \frac{\tau}{l+1} \left| \{ i = 1, \ldots, l+1 \mid \alpha_i^y = \alpha_{l+1}^y \} \right|, \qquad (2)$$

τ is a random number distributed uniformly on the interval $[0, 1]$ (even conditionally on all the examples), and the corresponding sequence of *conformity scores* is defined by

$$(\alpha_1^y, \ldots, \alpha_l^y, \alpha_{l+1}^y) := A(z_1, \ldots, z_l, (x, y)).$$

Notice that the system of *prediction sets* (1) output by a conformal predictor is decreasing in ϵ, or *nested*.

The *conformal transducer* determined by A outputs the system of p-values $(p^y \mid y \in \mathbf{Y})$ defined by (2) for each training sequence $(z_1, \ldots, z_l)$ of examples and each test object x. (This is just a different representation of the conformal predictor.)

The standard property of validity for conformal predictors and transducers is that the p-values p^y are distributed uniformly on $[0, 1]$ when the examples $z_1, \ldots, z_l, (x, y)$ are generated independently from the same probability distribution Q on $\mathbf{Z}$ (see, e.g., [14], Proposition 2.8). This implies that the probability of error, $y \notin \Gamma^\epsilon(z_1, \ldots, z_l, x)$, is ϵ at any significance level ϵ.

Suppose we are given a test sequence $(z_{l+1}, \ldots, z_{l+k})$ and would like to use it to measure the efficiency of the predictions derived from the training sequence $(z_1, \ldots, z_l)$. (The efficiency of conformal predictors means that the prediction sets they output tend to be small, and the efficiency of conformal transducers means that the p-values that they output tend to be small.) For each test example $z_i = (x_i, y_i)$, $i = l+1, \ldots, l+k$, we have a nested family $(\Gamma_i^\epsilon \mid \epsilon \in (0, 1))$ of subsets of $\mathbf{Y}$ and a system of p-values $(p_i^y \mid y \in \mathbf{Y})$. In this paper we will discuss ten criteria of efficiency for such a family or a system, but some of them will depend, additionally, on the observed labels y_i of the test examples. We start from the *prior* criteria, which do not depend on the observed test labels.

2.1   Basic Criteria

We will discuss two kinds of criteria: those applicable to the prediction sets Γ_i^ϵ and so depending on the significance level ϵ and those applicable to systems of p-values $(p_i^y \mid y \in \mathbf{Y})$ and so independent of ϵ. The simplest criteria of efficiency are:

– The S *criterion* (with "S" standing for "sum") measures efficiency by the average sum

$$\frac{1}{k} \sum_{i=l+1}^{l+k} \sum_y p_i^y \qquad (3)$$

of the p-values; small values are preferable for this criterion. It is ϵ-free.

- The *N criterion* uses the average size

$$\frac{1}{k} \sum_{i=l+1}^{l+k} |\Gamma_i^\epsilon|$$

of the prediction sets ("N" stands for "number": the size of a prediction set is the number of labels in it). Small values are preferable. Under this criterion the efficiency is a function of the significance level ϵ.

Both these criteria are prior. The S criterion was introduced in [3] and the N criterion was introduced independently in [3,5], although the analogue of the N criterion for regression (where the size of a prediction set is defined to be its Lebesgue measure) had been used earlier in [9] (whose arXiv version was published in 2012).

2.2 Other Prior Criteria

A disadvantage of the basic criteria is that they look too stringent. Even for a very efficient conformal transducer, we cannot expect all p-values p^y to be small: the p-value corresponding to the true label will not be small with high probability; and even for a very efficient conformal predictor we cannot expect the size of its prediction set to be zero: with high probability it will contain the true label. The other prior criteria are less stringent. The ones that do not depend on the significance level are:

- The *U criterion* (with "U" standing for "unconfidence") uses the average unconfidence

$$\frac{1}{k} \sum_{i=l+1}^{l+k} \min_y \max_{y' \neq y} p_i^{y'} \tag{4}$$

over the test sequence, where the *unconfidence* for a test object x_i is the second largest p-value $\min_y \max_{y' \neq y} p_i^{y'}$; small values of (4) are preferable. The U criterion in this form was introduced in [3], but it is equivalent to using the average confidence (one minus unconfidence), which is very common.
If two conformal transducers have the same average unconfidence (which is presumably a rare event), the criterion compares the average credibilities

$$\frac{1}{k} \sum_{i=l+1}^{l+k} \max_y p_i^y \tag{5}$$

where the *credibility* for a test object x_i is the largest p-value $\max_y p_i^y$; smaller values of (5) are preferable. (Intuitively, a small credibility is a warning that the test object is unusual, and since such a warning presents useful information and the probability of a warning is guaranteed to be small, we want to be warned as often as possible.)

– The *F criterion* uses the average fuzziness

$$\frac{1}{k} \sum_{i=l+1}^{l+k} \left(\sum_y p_i^y - \max_y p_i^y \right), \tag{6}$$

where the *fuzziness* for a test object x_i is defined as the sum of all p-values apart from a largest one, i.e., as $\sum_y p_i^y - \max_y p_i^y$; smaller values of (6) are preferable. If two conformal transducers lead to the same average fuzziness, the criterion compares the average credibilities (5), with smaller values preferable.

Their counterparts depending on the significance level are:

– The *M criterion* uses the percentage of objects x_i in the test sequence for which the prediction set Γ_i^ϵ at significance level ϵ is *multiple*, i.e., contains more than one label. Smaller values are preferable. As a formula, the criterion prefers smaller

$$\frac{1}{k} \sum_{i=l+1}^{l+k} \mathbf{1}_{\{|\Gamma_i^\epsilon|>1\}}, \tag{7}$$

where $\mathbf{1}_E$ denotes the indicator function of the event E (taking value 1 if E happens and 0 if not). When the percentage (7) of multiple predictions is the same for two conformal predictors (which is a common situation: the percentage can well be zero), the M criterion compares the percentages

$$\frac{1}{k} \sum_{i=l+1}^{l+k} \mathbf{1}_{\{\Gamma_i^\epsilon=\emptyset\}} \tag{8}$$

of empty predictions (larger values are preferable). This is a widely used criterion. (In particular, it was used in [14] and papers preceding it.)
– The *E criterion* (where "E" stands for "excess") uses the average (over the test sequence, as usual) amount the size of the prediction set exceeds 1. In other words, the criterion gives the average number of excess labels in the prediction sets as compared with the ideal situation of one-element prediction sets. Smaller values are preferable for this criterion. As a formula, the criterion prefers smaller

$$\frac{1}{k} \sum_{i=l+1}^{l+k} \left(|\Gamma_i^\epsilon| - 1 \right)^+,$$

where $t^+ := \max(t, 0)$. When these averages coincide for two conformal predictors, we compare the percentages (8) of empty predictions; larger values are preferable.

2.3   Observed Criteria

The prior criteria discussed in the previous subsection treat the largest p-value, or prediction sets of size 1, in a special way. The corresponding criteria of this subsection attempt to achieve the same goal by using the observed label.

These are the observed counterparts of the non-basic prior ϵ-free criteria:

- The OU ("observed unconfidence") *criterion* uses the average observed unconfidence

$$\frac{1}{k} \sum_{i=l+1}^{l+k} \max_{y \neq y_i} p_i^y$$

over the test sequence, where the *observed unconfidence* for a test example (x_i, y_i) is the largest p-value p_i^y for the *false labels* $y \neq y_i$. Smaller values are preferable for this test.

- The OF ("observed fuzziness") *criterion* uses the average sum of the p-values for the false labels, i.e.,

$$\frac{1}{k} \sum_{i=l+1}^{l+k} \sum_{y \neq y_i} p_i^y; \tag{9}$$

smaller values are preferable.

The counterparts of the last group depending on the significance level ϵ are:

- The OM *criterion* uses the percentage of observed multiple predictions

$$\frac{1}{k} \sum_{i=l+1}^{l+k} \mathbf{1}_{\{\Gamma_i^\epsilon \setminus \{y_i\} \neq \emptyset\}}$$

in the test sequence, where an *observed multiple* prediction is defined to be a prediction set including a false label. Smaller values are preferable.

- The OE *criterion* (OE standing for "observed excess") uses the average number

$$\frac{1}{k} \sum_{i=l+1}^{l+k} |\Gamma_i^\epsilon \setminus \{y_i\}|$$

of false labels included in the prediction sets at significance level ϵ; smaller values are preferable.

Table 1. The ten criteria studied in this paper: the two basic ones in the upper section; the four other prior ones in the middle section; and the four observed ones in the lower section

ϵ-free	ϵ-dependent
S (sum of p-values)	N (number of labels)
U (unconfidence)	M (multiple)
F (fuzziness)	E (excess)
OU (observed unconfidence)	OM (observed multiple)
OF (observed fuzziness)	OE (observed excess)

The ten criteria used in this paper are given in Table 1. Half of the criteria depend on the significance level ϵ, and the other half are the respective ϵ-free versions.

In the case of binary classification problems, $|\mathbf{Y}| = 2$, the number of different criteria of efficiency in Table 1 reduces to six: the criteria not separated by a vertical or horizontal line (namely, U and F, OU and OF, M and E, and OM and OE) coincide.

3   Optimal Idealised Conformity Measures for a Known Probability Distribution

Starting from this section we consider the limiting case of infinitely long training and test sequences (and we will return to the realistic finitary case only in Sect. 6, where we describe our empirical studies). To formalise the intuition of an infinitely long training sequence, we assume that the prediction algorithm is directly given the data-generating probability distribution Q on $\mathbf{Z}$ instead of being given a training sequence. Instead of conformity measures we will use *idealised conformity measures*: functions $A(Q, z)$ of $Q \in \mathcal{P}(\mathbf{Z})$ (where $\mathcal{P}(\mathbf{Z})$ is the set of all probability measures on $\mathbf{Z}$) and $z \in \mathbf{Z}$. We will fix the data-generating distribution Q for the rest of the paper, and so write the corresponding conformity scores as $A(z)$. The *idealised conformal predictor* corresponding to A outputs the following prediction set $\Gamma^\epsilon(x)$ for each object $x \in \mathbf{X}$ and each significance level $\epsilon \in (0, 1)$. For each potential label $y \in \mathbf{Y}$ for x define the corresponding *p-value* as

$$p^y = p(x, y) := Q\{z \in \mathbf{Z} \mid A(z) < A(x, y)\} + \tau Q\{z \in \mathbf{Z} \mid A(z) = A(x, y)\} \quad (10)$$

(it would be more correct to write $A((x, y))$ and $Q(\{\ldots\})$, but we often omit pairs of parentheses when there is no danger of ambiguity), where τ is a random number distributed uniformly on $[0, 1]$. (The same random number τ is used in (10) for all (x, y).) The prediction set is

$$\Gamma^\epsilon(x) := \{y \in \mathbf{Y} \mid p(x, y) > \epsilon\}. \quad (11)$$

The *idealised conformal transducer* corresponding to A outputs for each object $x \in \mathbf{X}$ the system of p-values $(p^y \mid y \in \mathbf{Y})$ defined by (10); in the idealised case we will usually use the alternative notation $p(x, y)$ for p^y.

The standard properties of validity for conformal transducers and predictors mentioned in the previous section simplify in this idealised case as follows:

- If (x, y) is generated from Q, $p(x, y)$ is distributed uniformly on $[0, 1]$.
- Therefore, at each significance level ϵ the idealised conformal predictor makes an error with probability ϵ.

The test sequence being infinitely long is formalised by replacing the use of a test sequence in the criteria of efficiency by averaging with respect to the data-generating probability distribution Q. In the case of the top two and bottom two criteria in Table 1 (the ones set in italics) this is done as follows. Let us write $\Gamma_A^\epsilon(x)$ for the $\Gamma^\epsilon(x)$ in (11) and $p_A(x,y)$ for the $p(x,y)$ in (10) to indicate the dependence on the choice of the idealised conformity measure A. An idealised conformity measure A is:

- *S-optimal* if, for any idealised conformity measure B,

$$\mathbb{E}_{x,\tau} \sum_{y \in \mathbf{Y}} p_A(x,y) \le \mathbb{E}_{x,\tau} \sum_{y \in \mathbf{Y}} p_B(x,y),$$

where the notation $\mathbb{E}_{x,\tau}$ refers to the expected value when x and τ are independent, $x \sim Q_{\mathbf{X}}$, and $\tau \sim U$; $Q_{\mathbf{X}}$ is the marginal distribution of Q on $\mathbf{X}$, and U is the uniform distribution on $[0,1]$;
- *N-optimal* if, for any idealised conformity measure B and any significance level ϵ,

$$\mathbb{E}_{x,\tau} |\Gamma_A^\epsilon(x)| \le \mathbb{E}_{x,\tau} |\Gamma_B^\epsilon(x)|;$$

- *OF-optimal* if, for any idealised conformity measure B,

$$\mathbb{E}_{(x,y),\tau} \sum_{y' \neq y} p_A(x,y') \le \mathbb{E}_{(x,y),\tau} \sum_{y' \neq y} p_B(x,y'),$$

where the lower index (x,y) in $\mathbb{E}_{(x,y),\tau}$ refers to averaging over $(x,y) \sim Q$ (with (x,y) and τ independent);
- *OE-optimal* if, for any idealised conformity measure B and any significance level ϵ,

$$\mathbb{E}_{(x,y),\tau} |\Gamma_A^\epsilon(x) \setminus \{y\}| \le \mathbb{E}_{(x,y),\tau} |\Gamma_B^\epsilon(x) \setminus \{y\}|.$$

We will define the idealised versions of the other six criteria listed in Table 1 in Sect. 5.

4   Probabilistic Criteria of Efficiency

Our goal in this section is to characterise the optimal idealised conformity measures for the four criteria of efficiency that are set in italics in Table 1. We will assume in the rest of the paper that the set $\mathbf{X}$ is finite (from the practical point of view, this is not a restriction); since we consider the case of classification, $|\mathbf{Y}| < \infty$, this implies that the whole example space $\mathbf{Z}$ is finite. Without loss of generality, we also assume that the data-generating probability distribution Q satisfies $Q_{\mathbf{X}}(x) > 0$ for all $x \in \mathbf{X}$ (we often omit curly braces in expressions such as $Q_{\mathbf{X}}(\{x\})$): we can always omit the xs for which $Q_{\mathbf{X}}(x) = 0$.

The *conditional probability (CP) idealised conformity measure* is

$$A(x,y) := Q(y \mid x) := \frac{Q(x,y)}{Q_{\mathbf{X}}(x)}. \tag{12}$$

This idealised conformity measure was introduced by an anonymous referee of the conference version of [3], but its non-idealised analogue in the case of regression had been used in [9] (following [8] and literature on minimum volume prediction). We say that an idealised conformity measure A is a *refinement* of an idealised conformity measure B if

$$B(z_1) < B(z_2) \implies A(z_1) < A(z_2) \tag{13}$$

for all $z_1, z_2 \in \mathbf{Z}$. Let $\mathcal{R}(\mathrm{CP})$ be the set of all refinements of the CP idealised conformity measure. If C is a criterion of efficiency (one of the ten criteria in Table 1), we let $\mathcal{O}(C)$ stand for the set of all C-optimal idealised conformity measures.

Theorem 1. $\mathcal{O}(\mathrm{S}) = \mathcal{O}(\mathrm{OF}) = \mathcal{O}(\mathrm{N}) = \mathcal{O}(\mathrm{OE}) = \mathcal{R}(\mathrm{CP})$.

We say that an efficiency criterion is *probabilistic* if the CP idealised conformity measure is optimal for it. Theorem 1 shows that four of our ten criteria are probabilistic, namely S, N, OF, and OE (they are set in italics in Table 1). In the next section we will see that in general the other six criteria are not probabilistic. The intuition behind probabilistic criteria will be briefly discussed also in the next section.

Proof (of Theorem 1). In this proof we partly follow [15], which simplified our original proof (considering, however, the case of label-conditional idealised conformal predictors and transducers).

We start from proving $\mathcal{R}(\mathrm{CP}) = \mathcal{O}(\mathrm{N})$. Let A be any idealised conformity measure. Fix for a moment a significance level ϵ. For each example $(x, y) \in \mathbf{Z}$, let $P(x, y)$ be the probability that the idealised conformal predictor based on A makes an error on the example (x, y) at the significance level ϵ, i.e., the probability of $y \notin \Gamma_A^\epsilon(x)$. It is clear from (10) and (11) that P takes at most three possible values (0, 1, and an intermediate value) and that

$$\sum_{x,y} Q(x, y) P(x, y) = \epsilon \tag{14}$$

(which just reflects the fact that the probability of error is ϵ). Vice versa, any P satisfying these properties will also satisfy

$$\forall (x, y) : P(x, y) = \mathbb{P}_{(x,y),\tau} \left(y \notin \Gamma_A^\epsilon(x) \right)$$

for some A. Let us see when we will have $A \in \mathcal{O}(\mathrm{N})$ (A is an N-optimal idealised conformity measure). Define Q' to be the probability measure on $\mathbf{Z}$ such that $Q'_{\mathbf{X}} = Q_{\mathbf{X}}$ and $Q'(y \mid x) = 1/|\mathbf{Y}|$ does not depend on y. The N criterion at significance level ϵ for A can be evaluated as

$$\mathbb{E}_{x,\tau} |\Gamma_A^\epsilon(x)| = |\mathbf{Y}| \left(1 - \sum_{(x,y)} Q'(x, y) P(x, y) \right);$$

this expression should be minimised, i.e., $\sum_{(x,y)} Q'(x,y)P(x,y)$ should be maximised, under the restriction (14). Let us apply the Neyman–Pearson fundamental lemma ([7], Sect. 3.2, Theorem 1) using Q as the null and Q' as the alternative hypotheses. We can see that $\mathbb{E}_{x,\tau} |\Gamma_A^\epsilon(x)|$ takes its maximal value if and only if there exist thresholds $k_1 = k_1(\epsilon)$, $k_2 = k_2(\epsilon)$, and $k_3 = k_3(\epsilon)$ such that:

- $Q\{(x,y) \mid Q(y \mid x) < k_1\} < \epsilon \le Q\{(x,y) \mid Q(y \mid x) \le k_1\}$,
- $k_2 < k_3$,
- $A(x,y) < k_2$ if $Q(y \mid x) < k_1$,
  $k_2 < A(x,y) < k_3$ if $Q(y \mid x) = k_1$,
- $A(x,y) > k_3$ if $Q(y \mid x) > k_1$.

This will be true for all ϵ if and only if $Q(y \mid x)$ is a function of $A(x,y)$ (meaning that there exists a function F such that, for all (x,y), $Q(y \mid x) = F(A(x,y))$). This completes the proof of $\mathcal{R}(\mathrm{CP}) = \mathcal{O}(\mathrm{N})$.

Next we show that $\mathcal{O}(\mathrm{N}) = \mathcal{O}(\mathrm{S})$. We will use the equality between the extreme terms of

$$\sum_{y \in \mathbf{Y}} p(x,y) = \sum_{y \in \mathbf{Y}} \int_0^1 \mathbf{1}_{\{p(x,y)>\epsilon\}} \, d\epsilon$$

$$= \int_0^1 \sum_{y \in \mathbf{Y}} \mathbf{1}_{\{p(x,y)>\epsilon\}} \, d\epsilon = \int_0^1 |\Gamma^\epsilon(x)| \, d\epsilon, \qquad (15)$$

which implies

$$\mathbb{E}_{x,\tau} \sum_{y \in \mathbf{Y}} p(x,y) = \int_0^1 \mathbb{E}_{x,\tau} |\Gamma^\epsilon(x)| \, d\epsilon. \qquad (16)$$

We can see that $A \in \mathcal{O}(\mathrm{S})$ whenever $A \in \mathcal{O}(\mathrm{N})$: indeed, any N-optimal idealised conformity measure minimises the expectation $\mathbb{E}_{x,\tau} |\Gamma^\epsilon(x)|$ on the right-hand side of (16) for all ϵ simultaneously, and so minimises the whole right-hand-side, and so minimises the left-hand-side. On the other hand, $A \notin \mathcal{O}(\mathrm{S})$ whenever $A \notin \mathcal{O}(\mathrm{N})$: indeed, if an idealised conformity measure fails to minimise the expectation $\mathbb{E}_{x,\tau} |\Gamma^\epsilon(x)|$ on the right-hand side of (16) for some ϵ, it fails to do so for all ϵ in a non-empty open interval (because of the right-continuity of $\mathbb{E}_{x,\tau} |\Gamma^\epsilon(x)|$ in ϵ, which follows from the Lebesgue dominated convergence theorem and the right-continuity of $|\Gamma^\epsilon(x)| = |\Gamma^\epsilon(x,\tau)|$ in ϵ for a fixed τ), and therefore, it does not minimise the right-hand side of (16) (any N-optimal idealised conformity measure, such as the CP idealised conformity measure, will give a smaller value), and therefore, it does not minimise the left-hand side of (16).

The equality $\mathcal{O}(\mathrm{S}) = \mathcal{O}(\mathrm{OF})$ follows from

$$\mathbb{E}_{x,\tau} \sum_y p(x,y) = \mathbb{E}_{(x,y),\tau} \sum_{y' \ne y} p(x,y') + \frac{1}{2},$$

where we have used the fact that $p(x,y)$ is distributed uniformly on $[0,1]$ when $((x,y),\tau) \sim Q \times U$ (see [14]).

Finally, we notice that $\mathcal{O}(\mathrm{N}) = \mathcal{O}(\mathrm{OE})$. Indeed, for any significance level ϵ,

$$\mathbb{E}_{x,\tau}\,|\Gamma^\epsilon(x)| = \mathbb{E}_{(x,y),\tau}\,|\Gamma^\epsilon(x) \setminus \{y\}| + (1 - \epsilon),$$

again using the fact that $p(x,y)$ is distributed uniformly on $[0,1]$ and so $\mathbb{P}_{(x,y),\tau}(y \in \Gamma^\epsilon(x)) = 1 - \epsilon$.   □

Remark 1. The statement $\mathcal{O}(\mathrm{S}) = \mathcal{R}(\mathrm{CP})$ of Theorem 1 can be generalised to the criterion S_ϕ preferring small values of

$$\frac{1}{k} \sum_{i=l+1}^{l+k} \sum_y \phi(p_i^y)$$

(instead of (3)), where $\phi : [0,1] \to \mathbb{R}$ is a fixed continuously differentiable strictly increasing function, not necessarily the identity function. Namely, we still have $\mathcal{O}(\mathrm{S}_\phi) = \mathcal{R}(\mathrm{CP})$. Indeed, we can assume, without loss of generality, that $\phi(0) = 0$ and $\phi(1) = 1$ and replace (15) by

$$\sum_{y \in \mathbf{Y}} \phi(p(x,y)) = \sum_{y \in \mathbf{Y}} \int_0^1 \mathbf{1}_{\{\phi(p(x,y)) > \epsilon\}}\, d\epsilon = \int_0^1 \sum_{y \in \mathbf{Y}} \mathbf{1}_{\{p(x,y) > \phi^{-1}(\epsilon)\}}\, d\epsilon$$

$$= \int_0^1 \left|\Gamma^{\phi^{-1}(\epsilon)}(x)\right|\, d\epsilon = \int_0^1 \left|\Gamma^{\epsilon'}(x)\right| \phi'(\epsilon')\, d\epsilon',$$

where ϕ' is the (continuous) derivative of ϕ, and then use the same argument as before.

5   Criteria of Efficiency that are Not Probabilistic

Now we define the idealised analogues of the six criteria that are not set in italics in Table 1. An idealised conformity measure A is:

- *U-optimal* if, for any idealised conformity measure B, we have either

$$\mathbb{E}_{x,\tau} \min_y \max_{y' \neq y} p_A(x, y') < \mathbb{E}_{x,\tau} \min_y \max_{y' \neq y} p_B(x, y')$$

or both

$$\mathbb{E}_{x,\tau} \min_y \max_{y' \neq y} p_A(x, y') = \mathbb{E}_{x,\tau} \min_y \max_{y' \neq y} p_B(x, y')$$

and

$$\mathbb{E}_{x,\tau} \max_y p_A(x, y) \leq \mathbb{E}_{x,\tau} \max_y p_B(x, y);$$

- *M-optimal* if, for any idealised conformity measure B and any significance level ϵ, we have either

$$\mathbb{P}_{x,\tau}(|\Gamma_A^\epsilon(x)| > 1) < \mathbb{P}_{x,\tau}(|\Gamma_B^\epsilon(x)| > 1)$$

or both

$$\mathbb{P}_{x,\tau}(|\Gamma_A^\epsilon(x)| > 1) = \mathbb{P}_{x,\tau}(|\Gamma_B^\epsilon(x)| > 1)$$

and

$$\mathbb{P}_{x,\tau}(|\Gamma_A^\epsilon(x)| = 0) \geq \mathbb{P}_{x,\tau}(|\Gamma_B^\epsilon(x)| = 0);$$

– *F-optimal* if, for any idealised conformity measure B, we have either

$$\mathbb{E}_{x,\tau}\left(\sum_y p_A(x,y) - \max_y p_A(x,y)\right) < \mathbb{E}_{x,\tau}\left(\sum_y p_B(x,y) - \max_y p_B(x,y)\right)$$

or both

$$\mathbb{E}_{x,\tau}\left(\sum_y p_A(x,y) - \max_y p_A(x,y)\right) = \mathbb{E}_{x,\tau}\left(\sum_y p_B(x,y) - \max_y p_B(x,y)\right)$$

and

$$\mathbb{E}_{x,\tau}\max_y p_A(x,y) \le \mathbb{E}_{x,\tau}\max_y p_B(x,y);$$

– *E-optimal* if, for any idealised conformity measure B and any significance level ϵ, we have either

$$\mathbb{E}_{x,\tau}\left((|\Gamma_A^\epsilon(x)| - 1)^+\right) < \mathbb{E}_{x,\tau}\left((|\Gamma_B^\epsilon(x)| - 1)^+\right)$$

or both

$$\mathbb{E}_{x,\tau}\left((|\Gamma_A^\epsilon(x)| - 1)^+\right) = \mathbb{E}_{x,\tau}\left((|\Gamma_B^\epsilon(x)| - 1)^+\right)$$

and

$$\mathbb{P}_{x,\tau}(|\Gamma_A^\epsilon(x)| = 0) \ge \mathbb{P}_{x,\tau}(|\Gamma_B^\epsilon(x)| = 0);$$

– *OU-optimal* if, for any idealised conformity measure B,

$$\mathbb{E}_{(x,y),\tau}\max_{y'\ne y} p_A(x,y') \le \mathbb{E}_{(x,y),\tau}\max_{y'\ne y} p_B(x,y');$$

– *OM-optimal* if, for any idealised conformity measure B and any significance level ϵ,

$$\mathbb{P}_{(x,y),\tau}(\Gamma_A^\epsilon(x) \setminus \{y\} \ne \emptyset) \le \mathbb{P}_{(x,y),\tau}(\Gamma_B^\epsilon(x) \setminus \{y\} \ne \emptyset).$$

In the following three definitions we follow [14], Chap. 3. The *predictability* of $x \in \mathbf{X}$ is

$$f(x) := \max_{y \in \mathbf{Y}} Q(y \mid x).$$

A *choice function* $\hat{y} : \mathbf{X} \to \mathbf{Y}$ is defined by the condition

$$\forall x \in \mathbf{X} : f(x) = Q(\hat{y}(x) \mid x).$$

Define the *signed predictability idealised conformity measure* corresponding to $\hat{y}$ by

$$A(x,y) := \begin{cases} f(x) & \text{if } y = \hat{y}(x) \\ -f(x) & \text{if not}; \end{cases}$$

a *signed predictability (SP) idealised conformity measure* is the signed predictability idealised conformity measure corresponding to some choice function.

For the following two theorems we will need to modify the notion of refinement. Let $\mathcal{R}'(\text{SP})$ be the set of all idealised conformity measures A such that there exists an SP idealised conformity measure B that satisfies both (13) and

$$B(x,y_1) = B(x,y_2) \implies A(x,y_1) = A(x,y_2)$$

for all $x \in \mathbf{X}$ and $y_1, y_2 \in \mathbf{Y}$.

Theorem 2. $\mathcal{O}(\mathrm{U}) = \mathcal{O}(\mathrm{M}) = \mathcal{R}'(\mathrm{SP})$.

We omit the proofs of Theorems 2–4 in this version of the paper.

Define the *MCP (modified conditional probability) idealised conformity measure* corresponding to a choice function $\hat{y}$ by

$$A(x, y) := \begin{cases} Q(y \mid x) & \text{if } y = \hat{y}(x) \\ Q(y \mid x) - 1 & \text{if not;} \end{cases}$$

an *MCP idealised conformity measure* is an idealised conformity measure corresponding to some choice function; $\mathcal{R}'(\mathrm{MCP})$ is defined analogously to $\mathcal{R}'(\mathrm{SP})$ but using MCP rather than SP idealised conformity measures.

Theorem 3. $\mathcal{O}(\mathrm{F}) = \mathcal{O}(\mathrm{E}) = \mathcal{R}'(\mathrm{MCP})$.

The *modified signed predictability idealised conformity measure* is defined by

$$A(x, y) := \begin{cases} f(x) & \text{if } f(x) > 1/2 \text{ and } y = \hat{y}(x) \\ 0 & \text{if } f(x) \leq 1/2 \\ -f(x) & \text{if } f(x) > 1/2 \text{ and } y \neq \hat{y}(x), \end{cases}$$

where f is the predictability function; notice that this definition is unaffected by the choice of the choice function. Somewhat informally and assuming $|\mathbf{Y}| > 2$ (we are in the situation of Theorem 1 when $|\mathbf{Y}| = 2$), we define a set $\mathcal{R}''(\mathrm{MSP})$ in the same way as $\mathcal{R}'(\mathrm{MSP})$ (analogously to $\mathcal{R}'(\mathrm{SP})$) except that for $A \in \mathcal{R}''(\mathrm{MSP})$, $f(x) = 1/2$, and $y \neq \hat{y}(x)$ we allow $A(x, y) < A(x, \hat{y}(x))$.

Theorem 4. *If* $|\mathbf{Y}| > 2$, $\mathcal{O}(\mathrm{OU}) = \mathcal{O}(\mathrm{OM}) = \mathcal{R}''(\mathrm{MSP})$.

Theorems 2–4 show that the six criteria that are not set in italics in Table 1 are not probabilistic (except for OU and OM when $|\mathbf{Y}| = 2$, of course). Criteria of efficiency that are not probabilistic are somewhat analogous to "improper scoring rules" in probability forecasting (see, e.g., [2,4]). The optimal idealised conformity measures for the criteria of efficiency given in this paper that are not probabilistic have clear disadvantages, such as:

- They depend on the arbitrary choice of a choice function. In many cases there is a unique choice function, but the possibility of non-uniqueness is still awkward.
- They encourage "strategic behaviour" (such as ignoring the differences, which may be very substantial, between potential labels other than $\hat{y}(x)$ for a test object x when using the M criterion).

However, we do not use the terminology "proper/improper" in the case of criteria of efficiency for conformal prediction since it is conceivable that some non-probabilistic criteria of efficiency may turn out to be useful.

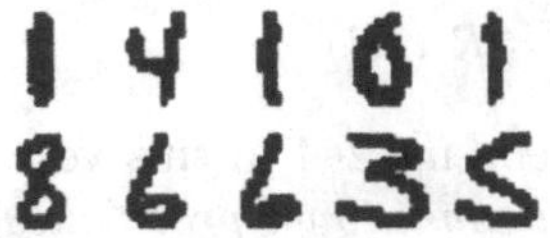

$$1\ 4\ 1\ 6\ 1$$
$$8\ 6\ 6\ 3\ 5$$

Fig. 1. Examples of hand-written digits in the USPS data set.

6 Empirical Study

In this section we demonstrate differences between two of our ϵ-free criteria, OF (probabilistic) and U (standard but not probabilistic) on the USPS data set of hand-written digits ([6]; examples of such digits are given in Fig. 1, which is a subset of Fig. 2 in [6]). We use the original split of the data set into the training and test sets. Our programs are written in R, and the results presented in the figures below are for the seed 0 of the R random number generator; however, we observe similar results in experiments with other seeds.

The problem is to classify hand-written digits, the labels are elements of $\{0, \ldots, 9\}$, and the objects are elements of $\mathbb{R}^{256}$, where the 256 numbers represent the brightness of pixels in 16×16 pictures. We normalise each object by applying the same affine transformation (depending on the object) to each of its pixels making the mean brightness of the pixels in the picture equal to 0 and making its standard deviation equal to 1. The sizes of the training and test sets are 7291 and 2007, respectively.

We evaluate six conformal predictors using the two criteria of efficiency. Fix a metric on the object space $\mathbb{R}^{256}$; in our experiments we use tangent distance (as implemented by Daniel Keysers) and Euclidean distance. Given a sequence of examples $(z_1, \ldots, z_n)$, $z_i = (x_i, y_i)$, we consider the following three ways of computing conformity scores: for $i = 1, \ldots, n$,

- $\alpha_i := \sum_{j=1}^{K} d_j^{\neq} / \sum_{j=1}^{K} d_j^{=}$, where $d_j^{\neq}$ are the distances, sorted in the increasing order, from x_i to the objects in $(z_1, \ldots, z_n)$ with labels different from y_i (so that $d_1^{\neq}$ is the smallest distance from x_i to an object x_j with $y_j \neq y_i$), and $d_j^{=}$ are the distances, sorted in the increasing order, from x_i to the objects in $(z_1, \ldots, z_{i-1}, z_{i+1}, \ldots, z_n)$ labelled as y_i (so that $d_1^{=}$ is the smallest distance from x_i to an object x_j with $j \neq i$ and $y_j = y_i$). We refer to this conformity measure as the *KNN-ratio conformity measure*; it has one parameter, K, whose range is $\{1, \ldots, 50\}$ in our experiments (so that we always have $K \ll n$).

- $\alpha_i := N_i / K$, where N_i is the number of objects labelled as y_i among the K nearest neighbours of x_i (when $d_K = d_{K+1}$ in the ordered list $d_1, \ldots, d_{n-1}$ of the distances from x_i to the other objects, we choose the nearest neighbours randomly among z_j with $y_j = y_i$ and with x_j at a distance of d_K from x_i). This conformity measure is a KNN counterpart of the CP idealised conformity measure (cf. (12)), and we will refer to it as the *KNN-CP conformity measure*; its parameter K is in the range $\{2, \ldots, 50\}$ in our experiments.

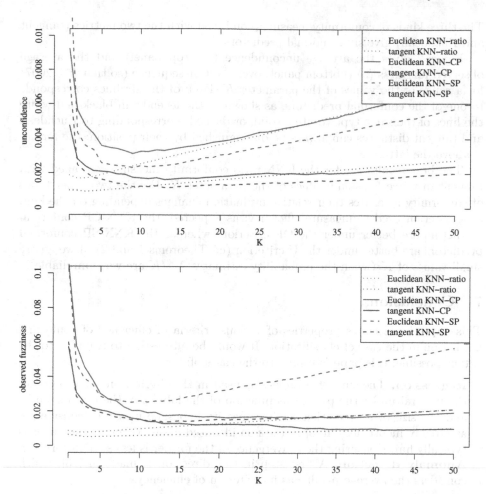

Fig. 2. Top plot: average unconfidence for the USPS data set (for different values of parameters). Bottom plot: average observed fuzziness for the USPS data set. In black-and-white the lines of the same type (dotted, solid, or dashed) corresponding to Euclidean and tangent distances can always be distinguished by their position: the former is above the latter.

– finally, we define $f_i := \max_y(N_i^y/K)$, where N_i^y is the number of objects labelled as y among the K nearest neighbours of x_i, $\hat{y}_i \in \arg\max_y(N_i^y/K)$ (chosen randomly from $\arg\max_y(N_i^y/K)$ if $|\arg\max_y(N_i^y/K)| > 1$), and

$$\alpha_i := \begin{cases} f_i & \text{if } y_i = \hat{y}_i \\ -f_i & \text{otherwise}; \end{cases}$$

this is the *KNN-SP conformity measure*.

The three kinds of conformity measures combined with the two metrics (tangent and Euclidean) give six conformal predictors.

Figure 2 gives the average unconfidence (4) (top panel) and the average observed fuzziness (9) (bottom panel) over the test sequence (so that $k = 2007$) for a range of the values of the parameter K. Each of the six lines corresponds to one of the conformal predictors, as shown in the legends; in black-and-white the lines of the same type (dotted, solid, or dashed) corresponding to Euclidean and tangent distances can always be distinguished by their position: the former is above the latter.

The best results are for the KNN-ratio conformity measure combined with tangent distance for small values of the parameter K. For the two other types of conformity measures their relative evaluation changes depending on the kind of a criterion used to measure efficiency: as expected, the KNN-CP conformal predictors are better under the OF criterion, whereas the KNN-SP conformal predictors are better under the U criterion (cf. Theorems 1 and 2), if we ignore small values of K (when the probability estimates N_i^y/K are very unreliable).

7   Conclusion

This paper investigates properties of various criteria of efficiency of conformal prediction in the case of classification. It would be interesting to transfer, to the extent possible, this paper's results to the cases of:

Regression. The sum of p-values (as used in the S criterion) now becomes the integral under the p-value as function of the label y of the text example, and the size of a prediction set becomes its Lebesgue measure (considered, as already mentioned, in [9] in the non-idealised case). Whereas the latter is typically finite, ensuring the convergence of the former is less straightforward.
Anomaly detection. A first step in this direction is made in [13], which considers the average p-value as its criterion of efficiency.

Other natural directions of further research include:

- Extensions of our results to infinite, including non-discrete, $\mathbf{X}$.
- Extensions to Mondrian conformal predictors. In the case of label-conditional conformal predictors and probabilistic criteria, this was started in [15].
- Extensions to non-idealised conformal predictors.

Acknowledgments. We are grateful to the reviewers for helpful comments. This work was partially supported by EPSRC (grant EP/K033344/1), the Air Force Office of Scientific Research (grant "Semantic Completions"), and the EU Horizon 2020 Research and Innovation programme (grant 671555).

References

1. Balasubramanian, V.N., Ho, S.S., Vovk, V. (eds.): Conformal Prediction for Reliable Machine Learning: Theory, Adaptations, and Applications. Elsevier, Amsterdam (2014)

2. Dawid, A.P.: Probability forecasting. In: Kotz, S., Balakrishnan, N., Read, C.B., Vidakovic, B., Johnson, N.L. (eds.) Encyclopedia of Statistical Sciences, vol. 10, 2nd edn, pp. 6445–6452. Wiley, Hoboken, NJ (2006)

3. Fedorova, V., Gammerman, A., Nouretdinov, I., Vovk, V.: Conformal prediction under hypergraphical models. In: Papadopoulos, H., Andreou, A.S., Iliadis, L., Maglogiannis, I. (eds.) AIAI 2013. IFIP AICT, vol. 412, pp. 371–383. Springer, Heidelberg (2013)

4. Gneiting, T., Raftery, A.E.: Strictly proper scoring rules, prediction, and estimation. J. Am. Stat. Assoc. **102**, 359–378 (2007)

5. Johansson, U., König, R., Löfström, T., Boström, H.: Evolved decision trees as conformal predictors. In: de la Fraga, L.G. (ed.) Proceedings of the 2013 IEEE Conference on Evolutionary Computation, vol. 1, pp. 794–1801. Cancun, Mexico (2013)

6. Le Cun, Y., Boser, B.E., Denker, J.S., Henderson, D., Howard, R.E., Hubbard, W.E., Jackel, L.D.: Handwritten digit recognition with a back-propagation network. In: Touretzky, D.S. (ed.) Advances in Neural Information Processing Systems 2, pp. 396–404. Morgan Kaufmann, San Francisco, CA (1990)

7. Lehmann, E.L.: Testing Statistical Hypotheses, 2nd edn. Springer, New York (1986)

8. Lei, J., Robins, J., Wasserman, L.: Distribution free prediction sets. J. Am. Stat. Assoc. **108**, 278–287 (2013)

9. Lei, J., Wasserman, L.: Distribution free prediction bands for nonparametric regression. J. Roy. Stat. Soc. B **76**, 71–96 (2014)

10. Melluish, T., Saunders, C., Nouretdinov, I., Vovk, V.: Comparing the Bayes and typicalness frameworks. In: Flach, P.A., De Raedt, L. (eds.) ECML 2001. LNCS (LNAI), vol. 2167, pp. 360–371. Springer, Heidelberg (2001)

11. Papadopoulos, H., Gammerman, A., Vovk, V.: Special issue of the conformal prediction and its applications. Ann. Math. Artif. Intell. **74**(1–2), 1–7 (2015). Springer

12. Saunders, C., Gammerman, A., Vovk, V.: Transduction with confidence and credibility. In: Dean, T. (ed.) Proceedings of the Sixteenth International Joint Conference on Artificial Intelligence, vol. 2, pp. 722–726. Morgan Kaufmann, San Francisco, CA (1999)

13. Smith, J., Nouretdinov, I., Craddock, R., Offer, C., Gammerman, A.: Anomaly detection of trajectories with kernel density estimation by conformal prediction. In: Iliadis, L., Maglogiannis, I., Papadopoulos, H., Sioutas, S., Makris, C. (eds.) Artificial Intelligence Applications and Innovations. IFIP AICT, vol. 437, pp. 271–280. Springer, Heidelberg (2014)

14. Vovk, V., Gammerman, A., Shafer, G.: Algorithmic Learning in a Random World. Springer, New York (2005)

15. Vovk, V., Petej, I., Fedorova, V.: From conformal to probabilistic prediction. In: Iliadis, L., Maglogiannis, I., Papadopoulos, H., Sioutas, S., Makris, C. (eds.) Artificial Intelligence Applications and Innovations. IFIP AICT, vol. 437, pp. 221–230. Springer, Heidelberg (2014)

Universal Probability-Free Conformal Prediction

Vladimir Vovk[1(✉)] and Dusko Pavlovic[2]

[1] Royal Holloway, University of London, Egham, Surrey, UK
v.vovk@rhul.ac.uk
[2] University of Hawaii, Honolulu, USA
dusko@hawaii.edu

Abstract. We construct a universal prediction system in the spirit of
Popper's falsifiability and Kolmogorov complexity. This prediction system does not depend on any statistical assumptions, but under the IID
assumption it dominates, although in a rather weak sense, conformal
prediction.

Keywords: Conformal prediction · Prediction systems · Probability-
free · Universal prediction

> Not for nothing do we call the laws of nature "laws":
> the more they prohibit, the more they say.

The Logic of Scientific Discovery
KARL POPPER

1 Introduction

In this paper we consider the problem of predicting labels, assumed to be binary,
of a sequence of objects. This is an online version of the standard problem
of binary classification. Namely, we will be interested in infinite sequences of
observations

$$\omega = (z_1, z_2, \ldots) = ((x_1, y_1), (x_2, y_2), \ldots) \in (\mathbf{X} \times 2)^\infty$$

(also called *infinite data sequences*), where $\mathbf{X}$ is an *object space* and $2 := \{0, 1\}$.
For simplicity, we will assume that $\mathbf{X}$ is a given finite set of, say, binary strings
(the intuition being that finite objects can always be encoded as binary strings).

Finite sequences $\sigma \in (\mathbf{X} \times 2)^*$ of observations will be called *finite data
sequences*. If σ_1, σ_2 are two finite data sequences, their concatenation will be
denoted (σ_1, σ_2); σ_2 is also allowed to be an element of $\mathbf{X} \times 2$. A standard partial order on $(\mathbf{X} \times 2)^*$ is defined as follows: $\sigma_1 \sqsubseteq \sigma_2$ means that σ_1 is a prefix of
σ_2; $\sigma_1 \sqsubset \sigma_2$ means that $\sigma_1 \sqsubseteq \sigma_2$ and $\sigma_1 \neq \sigma_2$.

We use the notation $\mathbb{N} := \{1, 2, \ldots\}$ for the set of positive integers and
$\mathbb{N}_0 := \{0, 1, 2, \ldots\}$ for the set of nonnegative integers. If $\omega \in (\mathbf{X} \times 2)^\infty$ and
$n \in \mathbb{N}_0$, $\omega^n \in (\mathbf{X} \times 2)^n$ is the prefix of ω of length n.

© Springer International Publishing Switzerland 2016
A. Gammerman et al. (Eds.): COPA 2016, LNAI 9653, pp. 40–47, 2016.
DOI: 10.1007/978-3-319-33395-3_3

A *situation* is a concatenation $(\sigma, x) \in (\mathbf{X} \times 2)^* \times \mathbf{X}$ of a finite data sequence σ and an object x; our task in the situation (σ, x) is to be able to predict the label of the new object x given the sequence σ of labelled objects. Given a situation $s = (\sigma, x)$ and a label $y \in 2$, we let (s, y) stand for the finite data sequence $(\sigma, (x, y))$, which is the concatenation of s and y.

2   Laws of Nature as Prediction Systems

According to Popper's [1] view of the philosophy of science, scientific laws of nature should be falsifiable: if a finite sequence of observations contradicts such a law, we should be able to detect it. (Popper often preferred to talk about scientific theories or statements instead of laws of nature.) The empirical content of a law of nature is the set of its potential falsifiers ([1], Sects. 31 and 35). We start from formalizing this notion in our toy setting, interpreting the requirement that we should be able to detect falsification as that we should be able to detect it eventually.

Formally, we define a *law of nature* L to be a recursively enumerable prefix-free subset of $(\mathbf{X} \times 2)^*$ (where *prefix-free* means that $\sigma_2 \notin L$ whenever $\sigma_1 \in L$ and $\sigma_1 \sqsubset \sigma_2$). Intuitively, these are the potential falsifiers, i.e., sequences of observations prohibited by the law of nature. The requirement of being recursively enumerable is implicit in the notion of a falsifier, and the requirement of being prefix-free reflects the fact that extensions of prohibited sequences of observations are automatically prohibited and there is no need to mention them in the definition.

A law of nature L gives rise to a prediction system: in a situation $s = (\sigma, x)$ it predicts that the label $y \in 2$ of the new object x will be an element of

$$\Pi_L(s) := \{y \in 2 \mid (s, y) \notin L\}. \tag{1}$$

There are three possibilities in each situation s:

- The law of nature makes a prediction, either 0 or 1, in situation s when the prediction set (1) is of size 1, $|\Pi_L(s)| = 1$.
- The prediction set is empty, $|\Pi_L(s)| = 0$, which means that the law of nature has been falsified.
- The law of nature refrains from making a prediction when $|\Pi_L(s)| = 2$. This can happen in two cases:
  - the law of nature was falsified in past: $\sigma' \in L$ for some $\sigma' \sqsubseteq \sigma$;
  - the law of nature has not been falsified as yet.

3   Strong Prediction Systems

The notion of a law of nature is static; experience tells us that laws of nature eventually fail and are replaced by other laws. Popper represented his picture of this process by formulas ("evolutionary schemas") similar to

$$\mathrm{PS}_1 \rightarrow \mathrm{TT}_1 \rightarrow \mathrm{EE}_1 \rightarrow \mathrm{PS}_2 \rightarrow \cdots \tag{2}$$

(introduced in his 1965 talk on which [2], Chap. 6, is based and also discussed in several other places in [2,3]; in our notation we follow Wikipedia). In response to a problem situation PS, a tentative theory TT is subjected to attempts at error elimination EE, whose success leads to a new problem situation PS and scientists come up with a new tentative theory TT, etc. In our toy version of this process, tentative theories are laws of nature, problem situations are situations in which our current law of nature becomes falsified, and there are no active attempts at error elimination (so that error elimination simply consists in waiting until the current law of nature becomes falsified).

If L and L' are laws of nature, we define $L \sqsubset L'$ to mean that for any $\sigma' \in L'$ there exists $\sigma \in L$ such that $\sigma \sqsubset \sigma'$. To formalize the philosophical picture (2), we define a *strong prediction system* $\mathcal{L}$ to be a nested sequence $L_1 \sqsubset L_2 \sqsubset \cdots$ of laws of nature $L_1, L_2, \ldots$ that are jointly recursively enumerable, in the sense of the set $\{(\sigma, n) \in (\mathbf{X} \times 2)^* \times \mathbb{N} \mid \sigma \in L_n\}$ being recursively enumerable.

The interpretation of a strong prediction system $\mathcal{L} = (L_1, L_2, \ldots)$ is that L_1 is the initial law of nature used for predicting the labels of new objects until it is falsified; as soon as it is falsified we start looking for and then using for prediction the following law of nature L_2 until it is falsified in its turn, etc. Therefore, the prediction set in a situation $s = (\sigma, x)$ is natural to define as the set

$$\Pi_{\mathcal{L}}(s) := \{y \in 2 \mid (s, y) \notin \cup_{n=1}^{\infty} L_n\}. \tag{3}$$

As before, it is possible that $\Pi_{\mathcal{L}}(s) = \emptyset$.

Fix a situation $s = (\sigma, x) \in (\mathbf{X} \times 2)^* \times \mathbf{X}$. Let $n = n(s)$ be the largest integer such that s has a prefix in L_n. It is possible that $n = 0$ (when s does not have such prefixes), but if $n \geq 1$, s will also have prefixes in $L_{n-1}, \ldots, L_1$, by the definition of a strong prediction system. Then L_{n+1} will be the current law of nature; all earlier laws, $L_n, L_{n-1}, \ldots, L_1$, have been falsified. The prediction (3) in situation s is then interpreted as the set of all observations y that are not prohibited by the current law L_{n+1}.

In the spirit of the theory of Kolmogorov complexity, we would like to have a universal prediction system. However, we are not aware of any useful notion of a universal strong prediction system. Therefore, in the next section we will introduce a wider notion of a prediction system that does not have this disadvantage.

4   Weak Prediction Systems and Universal Prediction

A *weak prediction system* $\mathcal{L}$ is defined to be a sequence (not required to be nested in any sense) $L_1, L_2, \ldots$ of laws of nature $L_n \subseteq (\mathbf{X} \times 2)^*$ that are jointly recursively enumerable.

Remark 1. Popper's evolutionary schema (2) was the simplest one that he considered; his more complicated ones, such as

$$\begin{array}{l} \nearrow TT_a \to EE_a \to PS_{2a} \to \cdots \\ PS_1 \to TT_b \to EE_b \to PS_{2b} \to \cdots \\ \searrow TT_c \to EE_c \to PS_{2c} \to \cdots \end{array}$$

(cf. [2], pp. 243 and 287), give rise to weak rather than strong prediction systems.

In the rest of this paper we will omit "weak" in "weak prediction system". The most basic way of using a prediction system $\mathcal{L}$ for making a prediction in situation $s = (\sigma, x)$ is as follows. Decide on the maximum number N of errors you are willing to make. Ignore all L_n apart from $L_1, \ldots, L_N$ in $\mathcal{L}$, so that the prediction set in situation s is

$$\Pi_{\mathcal{L}}^N(s) := \{y \in 2 \mid \forall n \in \{1, \ldots, N\} : (s, y) \notin L_n\}.$$

Notice that this way we are guaranteed to make at most N mistakes: making a mistake eliminates at least one law in the list $\{L_1, \ldots, L_N\}$.

Similarly to the usual theory of conformal prediction, another way of packaging $\mathcal{L}$'s prediction in situation s is, instead of choosing the threshold (or *level*) N in advance, to allow the user to apply her own threshold: in a situation s, for each $y \in 2$ report the attained level

$$\pi_{\mathcal{L}}^s(y) := \min \{n \in \mathbb{N} \mid (s, y) \in L_n\} \tag{4}$$

(with $\min \emptyset := \infty$). The user whose threshold is N will then consider $y \in 2$ with $\pi_{\mathcal{L}}^s(y) \leq N$ as prohibited in s. Notice that the function (4) is upper semicomputable (for a fixed $\mathcal{L}$).

The strength of a prediction system $\mathcal{L} = (L_1, L_2, \ldots)$ at level N is determined by its N-*part*

$$\mathcal{L}_{\leq N} := \bigcup_{n=1}^{N} L_n.$$

At level N, the prediction system L prohibits $y \in 2$ as continuation of a situation s if and only if $(s, y) \in \mathcal{L}_{\leq N}$.

The following lemma says that there exists a universal prediction system, in the sense that it is stronger than any other prediction system if we ignore a multiplicative increase in the number of errors made.

Lemma 1. *There is a* universal *prediction system* $\mathcal{U}$, *in the sense that for any prediction system* $\mathcal{L}$ *there exists a constant* $C > 0$ *such that, for any* N,

$$\mathcal{L}_{\leq N} \subseteq \mathcal{U}_{\leq CN}. \tag{5}$$

Proof. Let $\mathcal{L}^1, \mathcal{L}^2, \ldots$ be a recursive enumeration of all prediction systems; their component laws of nature will be denoted $(L_1^k, L_2^k, \ldots) := \mathcal{L}^k$. For each $n \in \mathbb{N}$, define the nth component U_n of $\mathcal{U} = (U_1, U_2, \ldots)$ as follows. Let the binary representation of n be

$$(a, 0, 1, \ldots, 1), \tag{6}$$

where a is a binary string (starting from 1) and the number of 1 s in the $1, \ldots, 1$ is $k - 1 \in \mathbb{N}_0$ (this sentence is the definition of $a = a(n)$ and $k = k(n)$ in terms of n). If the binary representation of n does not contain any 0s, a and k are undefined, and we set $U_n := \emptyset$. Otherwise, set

$$U_n := L_A^k,$$

where $A \in \mathbb{N}$ is the number whose binary representation is a. In other words, $\mathcal{U}$ consists of the components of $\mathcal{L}^k$, $k \in \mathbb{N}$; namely, L_1^k is placed in $\mathcal{U}$ as $U_{3 \times 2^{k-1} - 1}$ and then $L_2^k, L_3^k, \ldots$ are placed at intervals of 2^k:

$$U_{3 \times 2^{k-1} - 1 + 2^k (i-1)} = L_i^k, \quad i = 1, 2, \ldots.$$

It is easy to see that

$$\mathcal{L}_{\leq N}^k \subseteq \mathcal{U}_{\leq 3 \times 2^{k-1} - 1 + 2^k (N-1)}, \tag{7}$$

which is stronger than (5). □

Let us fix a universal prediction system $\mathcal{U}$. By $K(\mathcal{L})$ we will denote the smallest prefix complexity of the programs for computing a prediction system $\mathcal{L}$. The following lemma makes (5) uniform in $\mathcal{L}$ showing how C depends on $\mathcal{L}$.

Lemma 2. *There is a constant $C > 0$ such that, for any prediction system $\mathcal{L}$ and any N, the universal prediction system $\mathcal{U}$ satisfies*

$$\mathcal{L}_{\leq N} \subseteq \mathcal{U}_{\leq C 2^{K(\mathcal{L})} N}. \tag{8}$$

Proof. Follow the proof of Lemma 1 replacing the "code" $(0, 1, \ldots, 1)$ for $\mathcal{L}^k$ in (6) by any prefix-free description of $\mathcal{L}^k$ (with its bits written in the reverse order). Then the modification

$$\mathcal{L}_{\leq N}^k \subseteq \mathcal{U}_{\leq 2^{k'+1} - 1 + 2^{k'} (N-1)}$$

of (7) with $k' := K(\mathcal{L}^k)$ implies that (8) holds for some universal prediction system, which, when combined with the statement of Lemma 1, implies that (8) holds for our chosen universal prediction system $\mathcal{U}$. □

This is a corollary for laws of nature:

Corollary 1. *There is a constant C such that, for any law of nature L, the universal prediction system $\mathcal{U}$ satisfies*

$$L \subseteq \mathcal{U}_{\leq C 2^{K(L)}}. \tag{9}$$

Proof. We can regard laws of nature L to be a special case of prediction systems identifying L with $\mathcal{L} := (L, L, \ldots)$. It remains to apply Lemma 2 to $\mathcal{L}$ setting $N := 1$. □

We can equivalently rewrite (5), (8), and (9) as

$$\Pi_{\mathcal{U}}^{CN}(s) \subseteq \Pi_{\mathcal{L}}^{N}(s), \tag{10}$$

$$\Pi_{\mathcal{U}}^{C2^{K(\mathcal{L})}N}(s) \subseteq \Pi_{\mathcal{L}}^{N}(s), \tag{11}$$

and

$$\Pi_{\mathcal{U}}^{C2^{K(L)}}(s) \subseteq \Pi_{L}(s), \tag{12}$$

respectively, for all situations s. Intuitively, (10) says that the prediction sets output by the universal prediction system are at least as precise as the prediction sets output by any other prediction system $\mathcal{L}$ if we ignore a constant factor in specifying the level N; and (11) and (12) indicate the dependence of the constant factor on $\mathcal{L}$.

5   Universal Conformal Prediction under the IID Assumption

Comparison of prediction systems and conformal predictors is hampered by the fact that the latter are designed for the case where we have a constant amount of noise for each observation, and so we expect the number of errors to grow linearly rather than staying bounded. In this situation a reasonable prediction set is $\Pi_{\mathcal{L}}^{\epsilon N}(s)$, where N is the number of observations in the situation s. For a small ϵ using $\Pi_{\mathcal{L}}^{\epsilon N}(s)$ means that we trust the prediction system whose percentage of errors so far is at most ϵ.

Up to this point our exposition has been completely probability-free, but in the rest of this section we will consider the special case where the data are generated in the IID manner. For simplicity, we will only consider computable conformity measures that take values in the set $\mathbb{Q}$ of rational numbers.

Corollary 2. *Let Γ be a conformal predictor based on a computable conformity measure taking values in $\mathbb{Q}$. Then there exists $C > 0$ such that, for almost all infinite sequences of observations $\omega = ((x_1, y_1), (x_2, y_2), \ldots) \in (\mathbf{X} \times 2)^{\infty}$ and all significance levels $\epsilon \in (0,1)$, from some N on we will have*

$$\Pi_{\mathcal{U}}^{CN\epsilon \ln^2(1+1/\epsilon)}((\omega^N, x_{N+1})) \subseteq \Gamma^{\epsilon}((\omega^N, x_{N+1})). \tag{13}$$

This corollary asserts that the prediction set output by the universal prediction system is at least as precise as the prediction set output by Γ if we increase slightly the significance level: from ϵ to $C\epsilon \ln^2(1 + 1/\epsilon)$. It involves not just multiplying by a constant (as is the case for (5) and (8)–(12)) but also the logarithmic term $\ln^2(1 + 1/\epsilon)$.

It is easy to see that we can replace the C in (13) by $C2^{K(\Gamma)}$, where C now does not depend on Γ (and $K(\Gamma)$ is the smallest prefix complexity of the programs for computing the conformity measure on which Γ is based).

Proof (of Corollary 2). Let
$$\epsilon' := 2^{\lceil \log \epsilon \rceil + 1},$$

where log stands for the base 2 logarithm. (Intuitively, we simplify ϵ, in the sense of Kolmogorov complexity, by replacing it by a number of the form 2^{-m} for an integer m, and make it at least twice as large as the original ϵ.) Define a prediction system (both weak and strong) $\mathcal{L}$ as, essentially, $\Gamma^{\epsilon'}$; formally, $\mathcal{L} := (L_1, L_2, \ldots)$ and L_n is defined to be the set of all ω^N, where ω ranges over the infinite data sequences and N over $\mathbb{N}$, such that the set

$$\left\{ i \in \{1, \ldots, N\} \mid y_i \notin \Gamma^{\epsilon'}((\omega^{i-1}, x_i)) \right\}$$

is of size n and contains N. The prediction system $\mathcal{L}$ is determined by ϵ', so that $K(\mathcal{L})$ does not exceed (apart from the usual additive constant) $K(\epsilon')$. By the standard validity property of conformal predictors ([6], Corollary 1.1), Hoeffding's inequality, and the Borel–Cantelli lemma,

$$\Pi_{\mathcal{L}}^{\epsilon' N}((\omega^N, x_{N+1})) \subseteq \Gamma^{\epsilon}((\omega^N, x_{N+1})) \tag{14}$$

from some N on almost surely. By Lemma 2 (in the form of (11)),

$$\Pi_{\mathcal{U}}^{C_1 2^{K(\epsilon')} \epsilon' N}((\omega^N, x_{N+1})) \subseteq \Pi_{\mathcal{L}}^{\epsilon' N}((\omega^N, x_{N+1})) \tag{15}$$

for all N. The statement (13) of the corollary is obtained by combining (14), (15), and

$$2^{K(\epsilon')} \leq C_2 \ln^2(1 + 1/\epsilon).$$

To check the last inequality, remember that $\epsilon' = 2^{-m}$ for an integer m, which we assume to be positive, without loss of generality; therefore, our task reduces to checking that

$$2^{K(m)} \leq C_3 \ln^2(1 + 2^m),$$

i.e.,

$$2^{K(m)} \leq C_4 m^2.$$

Since $2^{-K(m)}$ is the universal semimeasure on the positive integers (see, e.g., [5], Theorem 7.29), we even have

$$2^{K(m)} \leq C_5 m (\log m)(\log \log m) \cdots (\log \cdots \log m),$$

where the product contains all factors that are greater than 1 (see [4], Appendix A).   □

6   Conclusion

In this note we have ignored the computational resources, first of all, the required computation time and space (memory). Developing versions of our definitions and results taking into account the time of computations is a natural next step. In analogy with the theory of Kolmogorov complexity, we expect that the simplest and most elegant results will be obtained for computational models that are more flexible than Turing machines, such as Kolmogorov–Uspensky algorithms and Schönhage machines.

Acknowledgments. We thank the anonymous referees for helpful comments. This work has been supported by the Air Force Office of Scientific Research (grant "Semantic Completions"), EPSRC (grant EP/K033344/1), and the EU Horizon 2020 Research and Innovation programme (grant 671555).

References

1. Popper, K.R.: Logik der Forschung. Springer, Vienna (1934). English translation: The Logic of Scientific Discovery. Hutchinson, London (1959)
2. Popper, K.R.: Objective Knowledge: An Evolutionary Approach, revised edn. Clarendon Press, Oxford (1979). First edition: 1972
3. Popper, K.R.: All Life is Problem Solving. Routledge, Abingdon (1999)
4. Rissanen, J.: A universal prior for integers and estimation by minimum description length. Ann. Stat. **11**, 416–431 (1983)
5. Shen, A.: Around Kolmogorov complexity: basic notions and results. In: Vovk, V., Papadopoulos, H., Gammerman, A. (eds.) Measures of Complexity: Festschrift for Alexey Chervonenkis, pp. 75–115. Springer, Cham (2015). Chap. 7
6. Vovk, V.: The basic conformal prediction framework. In: Balasubramanian, V.N., Ho, S.S., Vovk, V. (eds.) Conformal Prediction for Reliable Machine Learning: Theory, Adaptations, and Applications, Chap. 1, pp. 3–19. Elsevier, Amsterdam (2014)

Applications of Conformal Prediction

Conformal Predictors for Compound Activity Prediction

Paolo Toccaceli[✉], Ilia Nouretdinov[✉], and Alexander Gammerman[✉]

Royal Holloway, University of London, Egham, UK
{paolo,ilia,alex}@cs.rhul.ac.uk
http://clrc.rhul.ac.uk/

Abstract. The paper presents an application of Conformal Predictors to a chemoinformatics problem of identifying activities of chemical compounds. The paper addresses some specific challenges of this domain: a large number of compounds (training examples), high-dimensionality of feature space, sparseness and a strong class imbalance. A variant of conformal predictors called Inductive Mondrian Conformal Predictor is applied to deal with these challenges. Results are presented for several non-conformity measures (NCM) extracted from underlying algorithms and different kernels. A number of performance measures are used in order to demonstrate the flexibility of Inductive Mondrian Conformal Predictors in dealing with such a complex set of data.

Keywords: Conformal prediction · Confidence estimation · Chemoinformatics · Non-conformity measure

1 Introduction

Compound Activity Prediction is one of the key research areas of Chemoinformatics. It is of critical interest for the pharmaceutical industry, as it promises to cut down the costs of the initial screening of compounds by reducing the number of lab tests needed to identify a bioactive compound. The focus is on providing a set of potentially active compounds that is significantly "enriched" in terms of prevalence of bioactive compounds compared to a purely random sample of the compounds under consideration. The paper is an extension of our work presented in [16].

While it is true that this objective in itself could be helped with the classical machine learning techniques that usually provide a bare prediction, the hedged predictions made by Conformal Predictors (CP) provide some additional information that can be used advantageously in a number of respects.

First, CPs will supply the valid measures of confidence in the prediction of bioactivities of the compounds. Second, they can provide prediction and confidence for individual compounds. Third, they can allow the ranking of compounds to optimize the experimental testing of given samples. Finally, the user can control the number of errors and other performance measures like precision and recall by setting up a required level of confidence in the prediction.

© Springer International Publishing Switzerland 2016
A. Gammerman et al. (Eds.): COPA 2016, LNAI 9653, pp. 51–66, 2016.
DOI: 10.1007/978-3-319-33395-3_4

2   Machine Learning Background

2.1   Conformal Predictors

Conformal Predictors described in [7,11,12] revolve around the notion of Conformity (or rather of Non-Conformity).

Intuitively, one way of viewing the problem of classification is to assign a label $\hat{y}$ to a new object x so that the example $(x, \hat{y})$ does not look out of place among the training examples $(x_1, y_1), (x_2, y_2), \ldots, (x_\ell, y_\ell)$. To find how "strange" the new example is in comparison with the training set, we use the Non-Conformity Measure (NCM) to measure $(x, \hat{y})$.

The advantage of approaching classification in this way is that this leads to a novel way to quantify the uncertainty of the prediction, under some rather general hypotheses.

A Non-Conformity Measure can be extracted from any machine learning algorithm, although there is no universal method to choose it. Note that we are not necessarily interested in the actual classification resulting from such "underlying" machine learning algorithm. What we are really interested in is an indication of how "unusual" an example appears, given a training set.

Armed with an NCM, it is possible to compute for any example (x, y) a p-value that reflects how good the new example from the test set fits (or conforms) with the training set. A more accurate and formal statement is: for a chosen $\epsilon \in [0, 1]$ it is possible to compute p-values for test objects so that they are (in the long run) smaller than ϵ with probability at most ϵ. Note that the key assumption here is that the examples in the training set and the test objects are *independent and identically distributed* (in fact, even a weaker requirement of *exchangeability* is sufficient).

The idea is then to compute for a test object a p-value of every possible choice of the label.

Once the p-values are computed, they can be put to use in one of the following ways:

- Given a significance level, ϵ, a *region predictor* outputs for each test object the set of labels (i.e., a region in the label space) such that the actual label is not in the set no more than a fraction ϵ of the times. If the prediction set consists of more than one label, the prediction is called *uncertain*, whereas if there are no labels in the prediction set, the prediction is *empty*.
- Alternatively, a *forced* prediction (chosen by the largest p-value) is given, alongside with its *credibility* (the largest p-value) and *confidence* (the complement to 1 of the second largest p-value).

2.2   Inductive Mondrian Conformal Predictors

In order to apply conformal predictors to both big and imbalanced datasets, we combine two variants of conformal predictors from [7,12]: Inductive (to reduce computational complexity) and Mondrian (to deal with imbalanced data sets) Conformal Predictors.

To combine the Mondrian Conformal Prediction with that of Inductive Conformal Prediction, we have to revise the definition of p-value for the Mondrian case so that it incorporates the changes brought about by splitting the training set and evaluating the α_i only in the calibration set.

It is customary to split the training set at index h so that examples with index $i \leq h$ constitute the proper training set and examples with index $i > h$ (and $i \leq \ell$) constitute the calibration set.

The p-values for a hypothesis $y_{\ell+1} = y$ about the label of $x_{\ell+1}$ are defined as

$$p(y) = \frac{|\{i = h+1, \ldots, \ell+1 : y_i = y, \alpha_i \geq \alpha_{\ell+1}\}|}{|\{i = h+1, \ldots, \ell+1 : y_i = y\}|}$$

In other words, the formula above considers only α_i associated with those examples in the calibration set that have the same label as that of the completion we are currently considering (note that also $\alpha_{\ell+1}$ is included in the set of α_i used for the comparison). As in the previous forms of p-value, the fraction of such α_i that are greater than or equal to $\alpha_{\ell+1}$ is the p-value.

Finally, it is important to note that Inductive Conformal Predictors can be applied under less restrictive conditions. The requirement of i.i.d. can in fact be dropped for the proper training set, as the i.i.d. property is relevant only for the populations on which we calculate and compare the α_i, that is, the calibration and testing set.

This will allow us to use NCM based on such methods as Cascade SVM (described in Sect. 3.1), including a stage of splitting big data into parts.

3   Application to Compound Activity Prediction

To evaluate the performance of CP for Compound Activity Prediction in a realistic scenario, we sourced the data sets from a public-domain repository of High Throughput assays, PubChem BioAssay [19].

The data sets on PubChem identify a compound with its CID (a unique compound identifier that can be used to access the chemical data of the compound in another PubChem database) and provide the result of the assay as Active/Inactive as well as providing the actual measurements on which the result was derived, e.g. viability (percentage of cells alive) of the sample after exposure to the compound.

To apply machine learning techniques to this problem, the compounds must be described in terms of a number of numerical attributes. There are several approaches to do this. The one that was followed in this study is to compute *signature descriptors* [6,13]. Each signature corresponds to the number of occurrences of a given labelled subgraph in the molecule graph, with subgraphs limited to those with a given depth. In this exercise the signature descriptors[1] had at

---

[1] The signature descriptors and other types of descriptors (e.g. circular descriptors) can be computed with the CDK Java package or any of its adaptations such as the RCDK package for the R statistical software.

most height 3. Examples can be found in [17]. The resulting data set is a sparse matrix of attributes (the signatures, on the columns) and examples (the compounds, on the rows).

We evaluated Conformal Predictors first with various underlying algorithms on the smallest of the data sets and then with various data sets using the underlying algorithm that performed best in previous set of tests.

3.1 Underlying Algorithms

As a first step in the study, we set out to extract relevant non-conformity measures from different underlying algorithms: Support Vector Machines (SVM), Nearest Neighbours, Naïve Bayes. The Non Conformity Measures for each of the three underlying algorithms are listed in Table 1.

Table 1. The non conformity measures for the three underlying algorithms

Underlying	Non conformity measure α_i	Comment
SVM	$-y_i d(x_i)$	(signed) distance from separating hyperplane
kNN	$\dfrac{\sum_{j \neq i: y_j = y_i}^{(k)} d(x_j, x_i)}{\sum_{j \neq i: y_j \neq y_i}^{(k)} d(x_j, x_i)}$	here the summation is on the k smallest values of $d(x_j, x_i)$
Naïve Bayes	$-\log p(y_i = c|x_i)$	p is the posterior probability estimated by Naïve Bayes

There are a number of considerations arising from the application of each of these algorithms to Compound Activity Prediction.

SVM. The usage of SVM in this domain poses a number of challenges. First of all, the number of training examples was large enough to create a problem for our computational resources. The scaling of SVM to large data sets is indeed an active research area [2,5,14,15], especially in the case of non-linear kernels[2]. We turned our attention to a simple approach proposed by Graf et al. [9], called Cascade SVM.

The sizes of the training sets considered here are too large to be handled comfortably by generally available SVM implementations, such as `libsvm` [4]. The approach we follow could be construed as a form of *training set editing*. Vapnik proved formally that it is possible to decompose the training into an n-ary tree of SVM trainings. The first layer of SVMs is trained on training sets obtained as a partition of the overall training set. Each SVM in the first layer outputs its set of support vectors (SVs) which is generally smaller than the training set. In the second layer, each SVM takes as training set the merging

---

[2] In the case of linear SVM, it is possible to tackle the formulation of the quadratic optimization problem at the heart of the SVM in the primal and solve it with techniques such as Stochastic Gradient Descent or L-BFGS, which lend themselves well to being distributed across an array of computational nodes.

of n of the SVs sets found in the first layer. Each layer requires fewer SVMs. The process is repeated until a layer requires only one SVM. The set of SVs emerging from the last layer is not necessarily the same that would be obtained by training on the whole set (but it is often a good approximation). If one wants to obtain that set, the whole training tree should be executed again, but this time the SVs obtained at the last layer would be merged into each of the initial training blocks. A new set of SVs would then be obtained at the end of the tree of SVMs. If this new set is the same as the one in the previous iteration, this is the desired set. If not, the process is repeated once more. In [9] it was proved that the process converges and that it converges to the same set of SVs that one would obtain by training on the whole training set in one go.

To give an intuitive justification, the fundamental observation is that the SVM decision function is entirely defined just by the Support Vectors. It is as if these examples contained all the information necessary for the classification. Moreover, if we had a training set composed only of the SVs, we would have obtained the same decision function. So, one might as well remove the non-SVs altogether from the training set.

In experiments discussed here, we followed a simplified approach. Instead of a tree of SVMs, we opted for a linear arrangement as shown in Fig. 1.

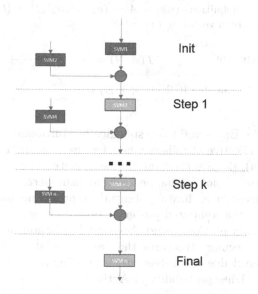

Fig. 1. Linear Cascade SVM. At each step, the set of Support Vectors from the previous stage is merged with a block of training examples from the partition of the original training set. This is used as training set for an SVM, whose SVs are then fed to the next stage.

While we have no theoretical support for this semi-online variant of the Cascade SVM, the method appears to work satisfactorily in practice on the data sets we used.

The class imbalance was addressed with the use of per-class weighting of the C hyperparameter, which results in a different penalization of the margin violations. The per-class weight was set inversely proportional to the class representation in the training set.

Another problem is the choice of an appropriate kernel. While we appreciated the computational advantages of linear SVM, we also believed that it was not necessarily the best choice for the specific problem. It can easily be observed that the nature of the representation of the training objects (as discrete features) warranted approaches similar to those used in Information Retrieval, where objects are described in terms of occurrences of patterns (bags of words). The topic of similarity searching in chemistry is an active one and there are many alternative proposals (see [1]). We used as a kernel a notion called Tanimoto similarity.[3] The Tanimoto similarity extends the well-known Jaccard coefficient in the sense that whereas the Jaccard coefficient considers only presence or absence of a pattern, the Tanimoto similarity takes into account the counts of the occurrences.

To explore further the benefits of non-linear kernels, we also tried out a kernel consisting of the composition the Tanimoto similarity with Gaussian RBF.

Table 2 provides the definitions of the kernels used in this study.

Table 2. SVM kernels definitions (where $A = (a_1, \ldots, a_d), B = (b_1, \ldots, b_d)$ are two objects, each described by a vector of d counts)

Tanimoto coefficient	$T(A, B) = \frac{\sum_{i=1}^{d} \min(a_i, b_i)}{\sum_{i=1}^{d}(a_i + b_i) - \sum_{i=1}^{d} \min(a_i, b_i)}$
Tanimoto with Gaussian RBF	$TG(A, B) = e^{-\frac{|T(A,A) + T(B,B) - 2T(A,B)|}{\gamma}}$

Naïve Bayes. Naïve Bayes and more specifically Multinomial Naïve Bayes are widely regarded as effective classifiers when features are discrete (for instance, in text classification), despite their relative simplicity. This made Multinomial Naïve Bayes a natural choice for the problem at issue here.

In addition, Naïve Bayes has a potential of providing some guidance for feature selection, via the computed posterior probabilities. This is of particular interest in the domain of Compound Activity Prediction, as it may provide insight as to the molecular structures that are associated with Activity in a given assay. This knowledge could steer further testing in the direction of a class of compounds with higher probability of Activity.

Nearest Neighbours. We chose Nearest Neighbours because of its good performance in a wide variety of domains. In principle, the performance of Nearest Neighbours could be severely affected by the high-dimensionality of the training set (Table 3 shows how in one of the data sets used in this study the number of attributes exceeds by $\approx 20\%$ the number of examples), but some preliminary small-scale experiments did not show that this causes the "curse of dimensionality".

---

[3] See [8] for a proof that Tanimoto Similarity is a kernel.

3.2   Tools and Computational Resources

The choice of the tools for these experiments was influenced primarily by the exploratory nature of this work. For this reason, tools, programming languages and environments that support interactivity and rapid prototyping were preferred to those that enable optimal CPU and memory efficiency.

The language adopted was Python 3.4 and the majority of programming was done using IPython Notebooks in the Jupyter environment. The overall format turned out to be very effective for capturing results (and for their future reproducibility).

Several third-party libraries were used. The computations were run initially on a local server (8 cores with 32GB of RAM, running OpenSuSE) and in later stages on a supercomputer (the IT4I Salomon cluster located in Ostrava, Czech Republic). The Salomon cluster is based on the SGI ICE X system and comprises 1008 computational nodes (plus a number of login nodes), each with 24 cores (2 12-core Intel Xeon E5-2680v3 2.5 GHz processors) and 128 GB RAM, connected via high-speed 7D Enhanced hypercube InfiniBand FDR and Ethernet networks. It currently ranks at #48 in the top500.org list of supercomputers and at #14 in Europe.[4]

Parallelization and computation distribution relied on the `ipyparallel` [3] package, which is a high-level framework for the coordination of remote execution of Python functions on a generic collection of nodes (cores or separate servers). While `ipyparallel` may not be highly optimized, it aims at providing a convenient environment for distributed computing well integrated with IPython and Jupyter and has a learning curve that is not as steep as that of the alternative frameworks common in High Performance Computing (OpenMPI, for example). In particular, `ipyparallel`, in addition to allowing the start-up and shut-down of a cluster comprising a controller and a number of engines where the actual processing (each is a separate process running a Python interpreter) is performed via integration with the job scheduling infrastructure present on Salomon (PBS, Portable Batch System), took care of the details such as data serialization/deserialization and transfer, load balancing, job tracking, exception propagation, etc. thereby hiding much of the complexity of parallelization. One key characteristic of `ipyparallel` is that, while it provides primitives for `map()` and `reduce()`, it does not constrain the choice to those two, leaving the implementer free to select the most appropriate parallel programming design patterns for the specific problem (see [20] for a reference on the subject).

In this work, parallelization was exploited to speed up the computation of the Gram matrix or of the decision function for the SVMs or the matrix of distances for kNN. In either case, the overall task was partitioned in smaller chunks that were then assigned to engines, which would then asynchronously return the result. Also, parallelization was used for SVM cross-validation, but at a coarser granularity, i.e. one engine per SVM training with a parameter. Data transfers were minimized by making use of shared memory where possible and appropriate. A key speed-up was

---

[4] According to https://www.sgi.com/company_info/newsroom/press_releases/2015/september/salomon.html.

achieved by using pre-computed kernels (computed once only) when performing Cross-Validation with respect to the hyperparameter C.

3.3 Results

To assess the relative merits of the different underlying algorithms, we applied Inductive Mondrian Conformal Predictors on data set AID827, whose characteristics are listed in Table 3.

Table 3. Characteristics of the AID827 data set

Total number of examples	138,287	
Number of features	165,786	High dimensionality
Number of non-zero entries	7,711,571	
Density of the data set	0.034 %	High sparsity
Active compounds	1,658	High imbalance (1.2 %)
Inactive compounds	136,629	
Unique set of signatures	137,901	Low degeneracy

The test was articulated in 20 cycles of training and evaluation. In each cycle, a test set of 10,000 examples was extracted at random. The remaining examples were split randomly into a proper training set of 100,000 examples and a calibration set with the balance of the examples (28,387).

During the SVM training, 5-fold stratified Cross Validation was performed at every stage of the Cascade to select an optimal value for the hyperparameter C. Also, per-class weights were assigned to cater for the high class imbalance in the data, so that a higher penalization was applied to violators in the less represented class.

In Multinomial Naïve Bayes too, Cross Validation was used to choose an optimal value for the smoothing parameter.

The results are listed in Table 4, which presents the classification arising from the region predictor for $\epsilon = 0.01$. The numbers are averages over the 20 cycles of training and testing.

Note that a compound is classified as Active (resp. Inactive) if and only if Active (resp. Inactive) is the only label in the prediction set. When both labels are in the prediction, the prediction is considered Uncertain.

It has to be noted at this stage that there does not seem to be an established consensus on what the best performance criteria are in the domain of Compound Activity Prediction (see for instance [10]), although *Precision* (fraction of actual Actives among compounds predicted as Active) and *Recall* (fraction of all the Active compounds that are among those predicted as Active) seem to be generally relevant. In addition, it is worth pointing out that these (and many others) criteria of performance should be considered as generalisations of classical performance criteria since they include dependence of the results on the required confidence level.

Table 4. CP results for AID827 with significance $\epsilon = 0.01$. All results are averages over 20 runs, using the same test sets of 10,000 objects across the different underlying algorithms. "Active predicted Active" is the (average) count of actually Active test examples that were predicted Active by Conformal Prediction. Uncertain predictions occur when both labels are output by the region predictor. Empty predictions occur when both labels can be rejected at the chosen significance level. For the specific significance level chosen here, there were never empty predictions.

Underlying	Active pred. Active	Inactive pred. Active	Inactive pred. Inactive	Active pred. Inactive	Empty pred.	Uncertain
Naïve Bayes	38.20	104.30	183.30	1.10	0	9673.10
3NN	43.95	100.55	361.55	0.80	0	9493.15
Cascade SVM						
- Linear	34.20	99.00	591.85	1.20	0	9273.75
- RBF kernel	47.20	101.80	1126.75	1.80	0	8722.45
- Tanimoto kernel	48.45	97.65	986.85	0.80	0	8866.25
- Tanimoto-RBF kernel	47.65	94.10	1044.90	0.95	0	8812.40

Table 5. CP results for AID827 using SVM with Tanimoto+RBF kernel for different significance levels. The "Active Error Rate" is the ratio of "Active predicted Inactive" to the total number of Active test examples. The "Inactive Error Rate" is the ratio of "Inactive predicted Active" to the total number of Inactive test examples.

Significance	Active pred. Active	Inactive pred. Active	Inactive pred. Inactive	Active pred. Inactive	Empty pred.	Uncertain	Active Error Rate	Inactive Error Rate
1 %	47.65	94.10	1044.90	0.95	0.0	8812.40	0.82 %	0.95 %
5 %	67.20	490.40	3091.75	5.20	0.0	6345.45	4.52 %	4.96 %
10 %	76.15	999.25	4703.75	10.60	0.0	4210.25	9.22 %	10.11 %
15 %	82.10	1484.85	6021.80	17.30	0.0	2393.95	15.04 %	15.02 %
20 %	86.55	1982.25	6928.95	22.80	0.0	979.45	19.83 %	20.05 %

At the shown significance level of $\epsilon = 0.01$, 34 % of the compounds predicted as active by Inductive Mondrian Conformal Prediction using Tanimoto composed with Gaussian RBF were actually Active compared to a prevalence of Actives in the data set of just 1.2 %. At the same time, the Recall was ≈ 41 % (ratio of Actives in the prediction to total Actives in the test set).

We selected Cascade SVM with Tanimoto+RBF as the most promising underlying algorithm on the basis of the combination of its high Recall (for Actives) and high Precision (for Actives), assuming that the intended application is indeed to output a selection of compounds that has a high prevalence of Active compounds.

Note that in Table 4 the values similar to ones of confusion matrix are calculated only for certain predictions. In this representation, the concrete meaning of the property of class-based validity can be clearly illustrated as in Table 5: the two rightmost columns report the prediction error rate for each label, where by prediction error we mean the occurrence of "the actual label not being in the predictions set". When there are no Empty predictions, the Active Error rate is the ratio of the number of "Active predicted Inactive" to the number of Active examples in the test set (which was 115 on average).

Figure 2 shows the test objects according to the base-10 logarithm of their p_{active} and $p_{inactive}$. The dashed lines represent the thresholds for p-value set at 0.01, i.e. the significance value ϵ used in Table 4. The two dashed lines partition the plane in 4 regions, corresponding to the region prediction being Active ($p_{active} > \epsilon$ and $p_{inactive} \leq \epsilon$), Inactive ($p_{active} \leq \epsilon$ and $p_{inactive} > \epsilon$), Empty ($p_{active} \leq \epsilon$ and $p_{inactive} \leq \epsilon$), Uncertain ($p_{active} > \epsilon$ and $p_{inactive} > \epsilon$).

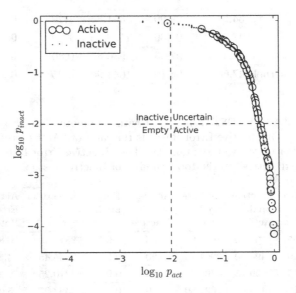

Fig. 2. Test objects plotted by the base-10 log of their p_{active} and $p_{inactive}$. Note that many test objects are overlapping. Note that some of the examples may have identical p-values, so for example 1135 objects predicted as "Inactives" are presented as 4 points on this plot.

As we said in Sect. 2.1, the alternative is forced prediction with individual confidence and credibility.

It is clear that there are several benefits accruing from using Conformal Predictors. For instance, a high p-value for the Active hypothesis might suggest that Activity cannot be ruled out, but the same compound may exhibit also a high p-value for the Inactive hypothesis, which would actually mean that neither hypothesis could be discounted.

In this specific context it can be argued that the p-values for Active hypothesis are more important. They can be used to rank the test compounds like it was done in [18] for ranking potential interaction. A high p-value for the Active hypothesis might suggest that Activity cannot be ruled out. For example it is possible to output the prediction list of all compounds with p-values above a threshold $\epsilon = 0.01$. A concrete activity which is not yet discovered will be covered by this list with probability 0.99. All the rest examples are classified as Non-Active with confidence 0.99 or larger.

Special attention should be also paid to *low credibility* examples where both p-values are small. Intuitively, low credibility means that either the training set is non-random or the test object is not representative of the training set. For such examples, the label assignment does not conform to the training data. They may be considered as anomalies or examples of compound types not enough represented in the training set. This may suggest that it would be beneficial to the overall performance of the classifier to perform a lab test for those compounds and include the results in training set. Typically credibility will not be low provided the data set was generated independently from the same distribution: the probability that credibility will be less than some threshold ϵ (such as 1 %) is less than ϵ.

Finally, Conformal Predictors provide the user with the additional degree of freedom of the significance or confidence level. By varying either of those two parameters, a different trade-off between Precision and Recall or any of the other metrics that are of interest can be chosen. Figure 3 illustrates this point with two examples. The Precision and Recall shown in the two panes were calculated on the test examples predicted Active which exceeded both a Credibility threshold and a given the Confidence threshold. In the left pane, the Credibility threshold was fixed and the Confidence threshold was varied; vice versa in the right pane.

3.4 Application to Different Data Sets

We applied Inductive Conformal Predictors with underlying SVM using Tanimoto+RBF kernel to other data sets extracted from PubChem BioAssay to verify if the same performance would be achieved for assays covering a range of

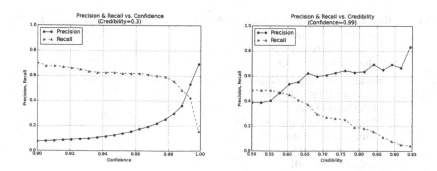

Fig. 3. Trade-off between Precision and Recall by varying credibility or confidence

quite different biological targets and to what extent the performance would vary with differences in training set size, imbalance, and sparseness of the training set. The main characteristics of the data sets are reported in Table 6.

As in the previous set of experiments, 20 cycles of training and testing were performed and the results averaged over them. In each cycle, a test set of 10,000 examples was set aside and the rest was split between calibration set ($\approx$30,000) and proper training set. The results are reported in Table 7.

It can be seen that five data sets differ in their hardness for machine learning some of the produce more uncertain predictions using the same algorithms, number of examples and the same significance level.

Table 6. Data sets and their characteristics. Density refers to the percentage of non-zero entries in the full matrix of 'Number of Compounds $\times$ Number of Features' elements

Data set	Assay description	Number of compounds	Number of features	Actives (%)	Density (%)
827	High throughput screen to identify compounds that suppress the growth of cells with a deletion of the PTEN tumor suppressor	138,287	165,786	1.2 %	0.034 %
1461	qHTS assay for antagonists of the neuropeptide S receptor: cAMP signal transduction	208,069	211,474	1.11 %	0.026 %
1974	Fluorescence polarization-based counterscreen for RBBP9 inhibitors: primary biochemical high throughput screening assay to identify inhibitors of the oxidoreductase glutathione S-transferase omega 1(GSTO1)	302,310	237,837	1.05 %	0.024 %
2553	High throughput screening of inhibitors of transient receptor potential cation channel C6 (TRPC6)	305,308	236,508	1.06 %	0.024 %
2716	Luminescence microorganism primary HTS to identify inhibitors of the SUMOylation pathway using a temperature sensitive growth reversal mutant Mot1-301	298,996	237,811	1.02 %	0.024 %

Table 7. Results of the application of Mondrian ICP with $\epsilon = 0.01$ using SVM with Tanimoto+RBF as underlying. Test set size: 10,000

DataSet	Active pred. Active	Inactive pred. Active	Inactive pred. Inactive	Active pred. Inactive	Empty pred.	Uncertain
827	47.65	94.10	1044.90	0.95	0	8812.40
1461	29.45	101.30	1891.10	1.20	0	7976.95
1974	62.50	97.40	880.85	1.00	0	8958.25
2553	34.00	101.00	337.90	1.00	0	9526.10
2716	3.55	98.20	97.00	1.00	0	9800.25

3.5 Mondrian ICP with Different ϵ_{active} and $\epsilon_{inactive}$

When applying Mondrian ICP, there is no constraint to use the same significance ϵ for the two labels. There may be an advantage in allowing different "error" rates for the two labels given that the focus might be in identifying Actives rather than Inactives.

This allows to vary relative importance of the two kinds of errors. Validity of Mondrian machines implies that the expected number of certain but wrong predictions is bounded by ϵ_{act} for (true) actives and by ϵ_{inact} for (true) non-actives. It is interesting to study its effect also on the precision and recall (within certain prediction).

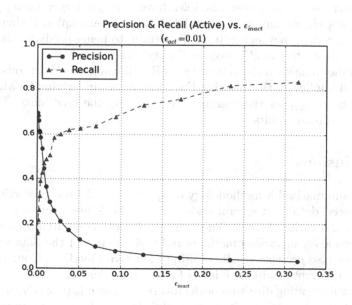

Fig. 4. Trade-off between Precision and Recall by varying ϵ_{inact}

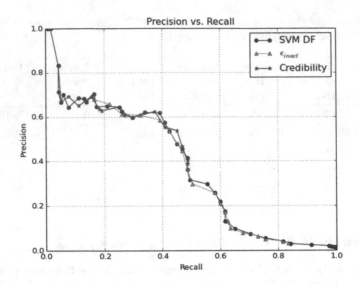

Fig. 5. Precision vs. Recall: three methods

Figure 4 shows the trade-off between Precision and Recall that results from varying ϵ_{inact}.

For very low values of the significance ϵ, a large number of test examples have $p_{act} > \epsilon_{act}$ as well as $p_{inact} > \epsilon_{inact}$. For these test examples, we have an 'Uncertain' prediction.

As we increase ϵ_{inact}, fewer examples have a p_{inact} larger than ϵ_{inact}. So 'Inactive' is not chosen any longer as a label for those examples. If they happen to have a $p_{act} > \epsilon_{act}$, they switch from 'Uncertain' to being predicted as 'Active' (in the other case, they would become 'Empty' predictions).

Figure 5 shows how Precision varies with Recall using three methods: varying the threshold applied to the Decision Function of the underlying SVM, varying the significance ϵ_{inact} for the Inactive class, varying the credibility. The three methods give similar results.

4   Conclusions

This paper summarized a methodology of applying conformal prediction to big and imbalanced data with several underlying methods like nearest neighbours, Bayes, SVM and various kernels. The results have been compared from the point of view of efficiency of various methods and various sizes of the data sets.

The paper also presents results of using Inductive Mondrian Conformal Predictors with different significance levels for different classes.

The most interesting direction of the future extension is to study the possible strategies of active learning (or experimental design). In this paper one of the criteria of performance is a number of uncertain predictions. It might be useful to select among them the compounds that should be checked experimentally first – in other words the most "promising" compounds. How to select though may depend on practical scenarios of further learning and on comparative efficiency of different active learning strategies.

Acknowledgments. This project (ExCAPE) has received funding from the European Unions Horizon 2020 Research and Innovation programme under Grant Agreement no. 671555. We are grateful for the help in conducting experiments to the Ministry of Education, Youth and Sports (Czech Republic) that supports the Large Infrastructures for Research, Experimental Development and Innovations project "IT4Innovations National Supercomputing Center LM2015070". This work was also supported by EPSRC grant EP/K033344/1 ("Mining the Network Behaviour of Bots"). We are indebted to Lars Carlsson of Astra Zeneca for providing the data and useful discussions. We are also thankful to Zhiyuan Luo and Vladimir Vovk for many valuable comments and discussions.

References

1. Monve, V.: Introduction to similarity searching in chemistry. MATCH - Comm. Math. Comp. Chem. **51**, 7–38 (2004)
2. Bottou, L., Chapelle, O., DeCoste, D., Weston, J.: Large-Scale Kernel Machines (Neural Information Processing). The MIT Press, Cambridge (2007)
3. Bussonnier, M.: Interactive parallel computing in Python. https://github.com/ipython/ipyparallel
4. Chang, C.-C., Lin, C.-J.: LIBSVM: a library for support vector machines. ACM Trans. Intell. Syst. Technol. **2**, 27:1–27:27 (2011). http://www.csie.ntu.edu.tw/~jlin/libsvm
5. Chang, E.Y.: PSVM: parallelizing support vector machines on distributed computers. Foundations of Large-Scale Multimedia Information Management and Retrieval, pp. 213–230. Springer, Heidelberg (2011)
6. Faulon Jr., J.-L., Visco, D.P., Pophale, R.S.: The signature molecular descriptor. 1. Using extended valence sequences in QSAR and QSPR studies. J. Chem. Inf. Comput. Sci. **43**(3), 707–720 (2003)
7. Gammerman, A., Vovk, V.: Hedging predictions in machine learning. Comput. J. **50**(2), 151–163 (2007)
8. Gärtner, T.: Kernels For Structured Data. World Scientific Publishing Co. Inc., River Edge (2009)
9. Graf, H.P., Cosatto, E., Bottou, L., Durdanovic, I., Vapnik, V.: Parallel Support Vector Machines: The Cascade SVM. In: Saul, L.K., Weiss, Y., Bottou, L. (eds.) Advances in Neural Information Processing Systems, pp. 521–528. MIT Press, Cambridge (2005)
10. Jain, A.N., Nicholls, A.: Recommendations for evaluation of computational methods. J. Comput. Aided Mol. Des. **22**(3–4), 133–139 (2008)
11. Shafer, G., Vovk, V.: A tutorial on conformal prediction. J. Mach. Learn. Res. **9**, 371–421 (2008)
12. Vovk, V., Gammerman, A., Shafer, G.: Algorithmic Learning in a Random World. Springer-Verlag New York, Inc., Secaucus (2005)

13. Weis, D.C., Visco Jr., D.P., Faulon, J.-L.: Data mining pubchem using a support vector machine with the signature molecular descriptor: classification of factor XIa inhibitors. J. Mol. Graph. Model. **27**(4), 466–475 (2008)
14. Woodsend, K., Gondzio, J.: Hybrid MPI/OpenMP parallel linear support vector machine training. J. Mach. Learn. Res. **10**, 1937–1953 (2009)
15. You, Y., Fu, H., Song, S.L., Randles, A., Kerbyson, D., Marquez, A., Yang, G., Hoisie, A.: Scaling support vector machines on modern HPC platforms. J. Parallel Distrib. Comput. **76**(C), 16–31 (2015)
16. Toccaceli, P., Nouretdinov, I., Luo, Z., Vovk, V., Carlsson, L., Gammerman, A.: Conformal predictors. Technical report for EU Horizon 2020 Programme ExCape Project. Royal Holloway, London, December 2015
17. Carlsson, L., Ahlberg, E., Boström, H., Johansson, U., Linusson, H.: Modifications to p-values of conformal predictors. In: SLDS 2015, pp. 251–259
18. Nouretdinov, I., Gammerman, A., Qi, Y., Klein-Seetharaman, J.: Determining confidence of predicted interactions between HIV-1 and human proteins using conformal method. In: Pacific Symposium on Biocomputing, p. 311 (2012)
19. Wang, Y., Suzek, T., Zhang, J., Wang, J., He, S., Cheng, T., Shoemaker, B.A., Gindulyte, A., Bryant, S.H.: PubChem BioAssay: 2014 update. Nucleic Acids Res. **42**(1), D1075–D1082 (2014)
20. McCool, M., Robison, A.D., Reinders, J.: Structured Parallel Programming: Patterns for Efficient Computation. Morgan-Kaufmann, Burlington (2012)

Conformal Prediction of Disruptions from Scratch: Application to an ITER Scenario

Raúl Moreno[1]([✉]), Jesús Vega[2], Sebastián Dormido[3], and JET Contributors[4]

[1] Departamento de Teoría de la Señal y Comunicaciones, UC3M, Madrid, Spain
raulms@tsc.uc3m.es
[2] Laboratorio Nacional de Fusión, CIEMAT, Madrid, Spain
[3] Departamento de Informática y Automática, UNED, Madrid, Spain
[4] EUROfusion Consortium, JET, Culham Science Centre,
Abingdon OX14 3DB, UK

Abstract. This article shows a conformal prediction application to disruption prediction from scratch. Considering data from ILW experimental campaigns (both hydrogen and deuterium campaigns), a one-layer disruption predictor has been tested from scratch. The results show a relevant improvement where the success rate (rate of disruptions predicted correctly) increases and the false alarm rate (rate of non-disruptive discharges misclassified) decreases, using conformal prediction (CP) rather than conventional methodology from scratch. CP from scratch achieves a success rate of 100 % with the first model and only one disruptive discharge.

Keywords: SVM · Conformal predictor · Disruption · Nuclear fusion

1 Introduction

Disruptions are plasma instabilities which lead the loss of the confinement and an abrupt drop of the plasma current [1,2]. The device is exposed to intense heat loads when the confinement is lost, hence, the first wall and plasma facing components can be seriously damaged by the high temperature and current. For this reason, disruptions are a serious problem at tokamak devices for the damage and the limit in the range of operation in current and density. Plasma can disrupt suddenly in a few milliseconds without precursors, or several plasma instabilities can finish disrupting in hundreds millisecond.

The concepts avoidance and mitigation are used to overcome disruptions. The first term is related to the scenario development and the plasma operational space free of disruptions [2,3], meanwhile the second term is referred to any action to reduce the harmful effect of disruptions and achieve a safe plasma shut-down [4-8]. Notwithstanding a successful avoidance and/or mitigation require an efficient prediction, it means, to have a reliable real-time disruption predictor during the experiments.

JET Contributors—See the Appendix of F. Romanelli et al., Proceedings of the 25th IAEA Fusion Energy Conference 2014, Saint Petersburg, Russia.

© Springer International Publishing Switzerland 2016
A. Gammerman et al. (Eds.): COPA 2016, LNAI 9653, pp. 67–74, 2016.
DOI: 10.1007/978-3-319-33395-3_5

The prediction of the incoming disruption with enough time to carry out mitigation actions plays an increasingly important role in the current and future devices. Nowadays, the largest nuclear fusion device is JET (Joint European Torus) until ITER (International Thermonuclear Experimental Reactor) operations begin. JET was operating with a Carbon fibre composite (CFC) wall until it was replaced by the new metallic ITER-like wall (ILW). Currently the closest approach to ITER is JET, unless the operational range is not the same. Although the development of more robust operational scenarios has reduced the JET disruption rate over the last decade from about 15–10 % to below 4 % [9]; disruptions probably will not be completely avoidable. ITER aims to operate with a disruption rate of 1 % or less. This rate is big enough to cause large damage on the device. To characterize plasma physics of disruption for their prediction is not a trivial task. The complexity of developing a physic driven system is due to the lack of theoretical knowledge on disruption phenomena, the large number of parameters involved in this stability and the non linear relation between them. Therefore, several data driven systems have been developed in the last 15 years, mainly based on neural networks and support vector machines (SVM), and they have been employed as an alternative approximation to detect the phenomena. Machine learning techniques are highly adaptable to disruptions because the instability can be considered as a classification problem. For this reason, the majority of research on disruption prediction has been based on data-driven models during the last years. Up to now, the data-driven architecture implemented in the real-time network of an experimental device that has provided the best detection rates has been APODIS at JET [10]. APODIS depend on a large database of past discharges to create models able to predict disruptions. However, next generation of tokamaks such as ITER and DEMO must be able to predict disruption from scratch, it means from the beginning of the operation with an absolutely lack of past discharges. Recently, two different works have dealt with the development of adaptive data-driven predictors from scratch to learn from the incoming data with good results: APODIS architecture from scratch [11] and probabilistic Venn's predictor [12].

Following this line of research, this paper provides an approach from scratch applying conformal prediction using a simple one-layer predictor reproducing a possible ITER scenario.

2   Conformal Prediction Review

A. Gammerman et al. introduced conformal prediction (CP) in [13], assigning values of confidence to predictions made by SVM. Then a complete theory on CP was developed where prediction algorithms (nearest-neighbor, SVM, ridge regression...) can be transformed into randomness tests and, therefore, be used for producing hedge predictions. Given an initial data set and an error probability ϵ, the new samples x_{n+1} are evaluated obtaining a prediction $\hat{y}$, which produces a set of labels y that also contains the label y_{n+1} with probability $1 - \epsilon$ [14]. Firstly, the concept of nonconformity measure is defined. It is a measure which represents how different is a new incoming sample from a bag of initial samples. Given a nonconformity measure A and a subset $Z_n = z_1, ..., z_n$, it can be calculated how different is a new sample z_{n+1} from Z_n. In [15], it can be seen that this nonconformity value is computed as $\alpha_{n+1} = A(Z_n, z_{n+1})$. However, the value of α_{n+1} does not tell how different is the sample z_{n+1}, so that it is necessary to compare α_{n+1} with the nonconformity values $\alpha_1, ..., \alpha_n$ from the samples of the subset Z_n. This comparison is called p-value:

$$\frac{\#i = 1, ..., n : \alpha_1 \geq \alpha n + 1}{n + 1}$$

This fraction is the p-value of the sample z_{n+1}. If this p-value is small, it means close to $1/n + 1$, then z_{n+1} is nonconforming (an outlier); while if the p-value is large, it means close to 1, then z_{n+1} is conforming. Nonconformity measures can be computed in several different ways, and each one defines a conformal predictor. In a classification problem with k classes, the p-value for the sample z_{n+1} has to be computed k times. Therefore, it is considered that sample z_{n+1} belongs to each one of the k classes and there will be k p-values. According to [16], the highest p-value, $P1$, determines the class predicted by the algorithm; and the second highest p-value, $P2$, defines the confidence in prediction: $P1$ is the credibility and $1 - P2$ is the confidence. The credibility serves as indicator of how suitable the training data are for classifying the example; while the confidence tells how likely each prediction is of being correct. This is the transductive CP approach, where the label of sample z_{n+1} is directly predicted using the training set. The computation is carried out for each test sample, and unfortunately it means a high computational cost. On the other hand, the inductive approach, extracts from the initial training set a general rule. This general rule is called model or decision rule and it is used to carry out the prediction of the new incoming samples. The CP is a transductive algorithm which can be highly inefficient for large data sets, so that there is an inductive CP approach. Nevertheless, a transductive approach is used in this work because disruption prediction from scratch begins the analysis with absolutely lack of previous information.

3   Databases and One-layer Predictor

This study considers a database formed by discharges (37 non-intentional disruptions and 110 non-disruptive discharges) from the JET hydrogen experimental campaign (September–October 2014) with the metallic ILW; and a database formed by discharges (351 non-intentional disruptions and 892 non-disruptive discharges) from the JET deuterium ILW experimental campaigns (July–September 2013). Intentional disruptions have been discarded from the database. Since they are programmed and forced to occur at a predefined instant of a discharge, plasma does not evolve naturally and their analysis may mislead a machine learning system.

Following previous studies [10,11,17], data have been processed calculating feature vectors which contain 2 values per signal: the standard deviation of the

Discrete Fourier Transform (discarding the DC component) computed over the past 32 ms of each signal, and the mean value of the amplitude calculated for that previous 32 ms. All signals have been processed following a strict real-time simulation, which means that it can be exactly reproduced under online conditions. The predictor chosen is the one-layer predictor obtained in [18]. This predictor is based on SVM with RBF kernel using libsvm in Matlab. This predictor uses only 3 signals (it means 6 features): plasma current, mode lock and plasma internal inductance. The disruptive samples considered correspond to the last 64 ms before the disruption time, it means, two disruptive samples per disruptive discharge $((64\,ms, 32\,ms], (32\,ms, Tdis])$. On the other hand, the non-disruptive samples selected are all samples from sample-23 until the last sample per safe discharge.

4   From Scratch Methodology

Following previous works [10,11], the problem to be solved is a binary classification but this methodology can be extrapolated to multi-class problems. The number of non-disruptive discharges is much higher than the number of disruptive discharges. In this sense, the fusion databases to develop disruption predictors are highly unbalanced. From scratch methodology trains models as discharges are produced, this means that discharges are used in chronological order as they occur. Regardless of the balanced or unbalanced data, it has to be established when a retraining is required to improve the predictor.

The first model is obtained after the first disruption and, from that moment, all discharges are analyzed in chronological order until the next missed alarm, it means, the next disruption which has not been predicted correctly. After a missed alarm, a new training set is created to incorporate new knowledge.

The approach selected in this work is unbalanced training data sets: it will be taken all the disruptive and non-disruptive discharges available until this moment. It is important to note that we are considering unbalanced number of discharges, the number of samples used in each disruptive and non-disruptive discharge follow the process explain previously.

5   ITER Scenario: One-layer Predictor from Scratch

In ITER views, the first operations will be carried out in hydrogen and helium because these plasmas have a lower power threshold to create the first ITER H-mode plasmas. At this point, an analysis from scratch, taking as the initial campaign the JET hydrogen campaign (September–October 2014) and then the rest ILW campaigns in chronological order (September 2011–September 2014), is developed to test a possible application to ITER.

As explained previously in Sects. 3 and 4, the one-layer predictor has been chosen to perform this analysis from scratch. Therefore, a first model is trained when the first disruption happens. This model is used until a new disruption is missed, at this point a new model is trained with all the previous data (all previous non-disruptive discharges and non-intentional disruptions). Hydrogen and deuterium campaigns have been tested from scratch separately. It can be seen in Fig. 1 the ILW deuterium campaigns results, 24 models have been developed (it means 24 missed alarms of 351 non-intentional disruptions) reaching a success rate of 93.43 % and a false alarm rate of 2.47 %. On the other hand, the performance on ILW hydrogen campaign, illustrated in Fig. 2, shows an erratic behavior training 4 models for 36 non-intentional disruptions, reaching a success rate of 91.67 % and false alarm rate of 60.91 %. Further analysis should be done to analyze the increment of false alarm rate, notwithstanding it can be understood as a consequence of the differences between hydrogen and deuterium experimental campaigns; and the less amount of data available on hydrogen campaign.

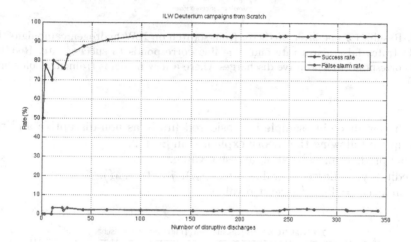

Fig. 1. Results for one-layer predictor from scratch on ILW deuterium campaigns. Red line illustrates false alarm rate and blue line corresponds to success rate. Red/Blue points correspond to disruptive discharges where a new model is trained. (Color figure online)

6   ITER Scenario: One-layer Conformal Predictor from Scratch

The previous section shows on hydrogen campaign from scratch a success rate of 91.67 % and false alarm rate of 60.91 %. The high false alarm rate means a bad performance; consequently, conformal predictor framework has been applied on this problem. The one-layer predictor is applied using transductive conformal prediction. The non-conformity measure will be the distance from the separating hyper plane, in the case of libsvm it will be the variable called decision values:

$$\alpha_i = \begin{cases} -|D(x_i)| & \text{if } x_i \text{ is well classified} \\ |D(x_i)| & \text{otherwise} \end{cases}$$

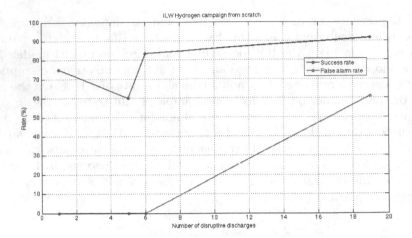

Fig. 2. Results for one-layer predictor from scratch on ILW hydrogen campaigns. Red line illustrates false alarm rate and blue line corresponds to success rate. Red/Blue points correspond to disruptive discharges where a new model is trained. (Color figure online)

Each new incoming sample is considered firstly as non-disruptive and then as disruptive following the theory explained in Sect. 2:

Label prediction: = label of largest p-value.

Credibility: = Largest p-value $(max(p_{yj}), j = 1, ..., M)$.

Confidence: = $1-2^{nd}$ largest p-value.

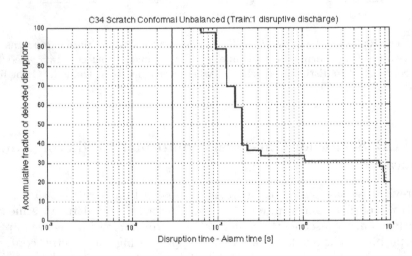

Fig. 3. Results for one-layer CP from scratch on ILW hydrogen campaigns. Green line represents 30 ms before the disruption time. It can be seen that 100 % of success rate is achieved with a warning time larger than 30 ms before disruption time (Color figure online).

In this framework, the one-layer CP only needs one model, it means only one disruptive discharge, to reach a success rate of 100 % (36/36) with a false alarm rate of 20 % (36/110), see Fig. 3. The 100 % of the cases are predicted with a warning time larger than 30 ms before disruption time, it can be seen clearly in Fig. 3, where the green line represents 30 ms before the disruption time. Despite the improvement on the results, the false alarm rate is still high. It is important to note that the methodology from scratch chosen in this study, retrains a model every missed alarm. Future work is proposed in the following section to minimize false alarm rate.

7   Discussion and Future Work

ITER cannot wait for several disruptions to have a reliable predictor. Furthermore, current ITER requirements aims a success rate $\geq$ 95 % and minimize the number of false alarms. This study provides a first ITER scenario on disruption prediction. The performance of one-layer predictor from scratch shows good results in terms of success rate and false alarm rate during deuterium campaigns. However, the performance on hydrogen campaigns reaches a good success rate but high false alarm rate. The first results using CP from scratch in disruption prediction show a notable improvement: the success rate increases until 100 % while false alarm decreases from 60 % to 20 % with no retraining. Only the first model is necessary under retraining conditions stablished: the system trains a new model every missed alarm. This study is being extended to the rest of the ILW campaigns in order to test the performance of CP from scratch with larger amount of data.

On the other hand, as mentioned previously in Sect. 6, CP from scratch on hydrogen campaigns obtains high success rate and reduces the false alarm rate from 60 % to 20 %. This rate is still high for ITER, however due to from scratch methodology chosen (a new model is trained after every missed alarm), the false alarm rate could be improved considering false alarms as retraining condition. For this reason, it is being studied the use of credibility and confidence as feature selection technique for disruptive and non-disruptive samples to improve the results.

Acknowledgments. This work was partially funded by the Spanish Ministry of Economy and Competitiveness under the Project No. ENE2012-38970-C04-01 and Project No. ENE2012-38970-C04-03.

This work has been carried out within the framework of the EUROfusion Consortium and has received funding from the Euratom research and training programme 2014–2018 under grant agreement No. 633053. The views and opinions expressed herein do not necessarily reflect those of the European Commission.

References

1. Wesson, J.A., Gill, R.D., Hugon, M., Schüller, F.C., Snipes, J.A., Ward, D.J., et al.: Disruptions in JET. Nucl. Fusion **29**(4), 641 (1989)
2. Hender, T.C., Wesley, J.C., Bialek, J., Bondeson, A., Boozer, A.H., Buttery, R.J., et al.: Chap. 3: MHD stability, operational limits and disruptions. Nucl. Fusion **47**(6), 128–202 (2007)
3. Esposito, B., Granucci, G., Nowak, S., Martin-Solis, J.R., Gabellieri, L., Lazzaro, E., et al.: Disruption control on FTU and ASDEX upgrade with ECRH. Nucl. Fusion **49**(6), 065014 (2009)
4. Reux, C., Bucalossi, J., Saint-Laurent, F., Gil, C., Moreau, P., Maget, P.: Experimental study of disruption mitigation using massive injection of noble gases on Tore Supra. Nucl. Fusion **50**(9), 095006 (2010)
5. Commaux, N., Baylor, L.R., Jernigan, T.C., Hollmann, E.M., Parks, P.B., Humphreys, D.A., et al.: Demonstration of rapid shutdown using large shattered deuterium pellet injection in DIII-D. Nucl. Fusion **50**(11), 112001 (2010)
6. Bakhtiari, M., Olynyk, G., Granetz, R., Whyte, D.G., Reinke, M.L., Izzo, V.: Using mixed gases for massive gas injection disruption mitigation on Alcator C-Mod. Nucl. Fusion **51**(6), 063007 (2011)
7. Commaux, N., Baylor, L.R., Combs, S.K., Eidietis, N.W., Evans, T.E., Foust, C.R., et al.: Novel rapid shutdown strategies for runaway electron suppression in DIII-D. Nucl. Fusion **51**(10), 103001 (2011)
8. Lehnen, M., Alonso, A., Arnoux, G., Baumgarten, N., Bozhenkov, S.A., Brezinsek, S., et al.: Disruption mitigation by massive gas injection in JET. Nucl. Fusion **51**(12), 123010 (2011)
9. de Vries, P.C., Johnson, M.F., Segui, I.: Statistical analysis of disruptions in JET. Nucl. Fusion **49**(5), 055011 (2009)
10. Vega, J., Dormido-Canto, S., López, J.M., Murari, A., Ramírez, J.M., Moreno, R., et al.: Results of the JET real-time disruption predictor in the ITER-like wall campaigns. Fusion Eng. Des. **88**(6–8), 1228–1231 (2013)
11. Dormido-Canto, S., Vega, J., Ramírez, J.M., Murari, A., Moreno, R., López, J.M., et al.: Development of an efficient real-time disruption predictor from scratch on JET and implications for ITER. Nucl. Fusion **53**(11), 113001 (2013)
12. Vega, J., Murari, A., Dormido-Canto, S., Moreno, R., Pereira, A., Acero, A.: Adaptive high learning rate probabilistic disruption predictors from scratch for the next generation of tokamaks. Nucl. Fusion **54**(12), 123001 (2014)
13. Gammerman, A., Vovk, V., Vapnik, V.: Learning by transduction. In: Proceedings of the Fourteenth Conference on Uncertainty in Artificial Intelligence, Madison, Wisconsin. 2074112, pp. 148–55. Morgan Kaufmann Publishers Inc. (1998)
14. Shafer, G., Vovk, V.: A tutorial on conformal prediction. J. Mach. Learn. Res. **9**, 371–421 (2008)
15. Vovk, V., Gammerman, A., Shafer, G.: Algorithmic Learning in a Random World. Springer, Berlin (2005)
16. Saunders, C., Gammerman, A., Vovk, V.: Transduction with Confidence and Credibility, pp. 722–726. Morgan Kaufmann Pub Inc., San Francisco (1999)
17. Rattá, G.A., Vega, J., Murari, A., Johnson, M.: Feature extraction for improved disruption prediction analysis at JET^a. Rev. Sci. Instrum. **79**(10), 10F328 (2008)
18. Moreno, R., Vega, J., Dormido-Canto, S., Pereira, A., Murari, A.: Disruption prediction on JET during the ILW experimental campaigns. Fus. Sci. Technol. **69**(2) (2016). doi:10.13182/FST15-167

Evaluation of a Variance-Based Nonconformity Measure for Regression Forests

Henrik Boström[1]([✉]), Henrik Linusson[2], Tuve Löfström[2], and Ulf Johansson[3]

[1] Department of Computer and Systems Sciences, Stockholm University,
Kista, Sweden
henrik.bostrom@dsv.su.se
[2] Department of Information Technology, University of Borås, Borås, Sweden
{henrik.linusson,tuve.lofstrom}@hb.se
[3] Department of Computer Science and Informatics, Jönköping University,
Jönköping, Sweden
ulf.johansson@ju.se

Abstract. In a previous large-scale empirical evaluation of conformal regression approaches, random forests using out-of-bag instances for calibration together with a k-nearest neighbor-based nonconformity measure, was shown to obtain state-of-the-art performance with respect to efficiency, i.e., average size of prediction regions. However, the use of the nearest-neighbor procedure not only requires that all training data have to be retained in conjunction with the underlying model, but also that a significant computational overhead is incurred, during both training and testing. In this study, a more straightforward nonconformity measure is investigated, where the difficulty estimate employed for normalization is based on the variance of the predictions made by the trees in a forest. A large-scale empirical evaluation is presented, showing that both the nearest-neighbor-based and the variance-based measures significantly outperform a standard (non-normalized) nonconformity measure, while no significant difference in efficiency between the two normalized approaches is observed. Moreover, the evaluation shows that state-of-the-art performance is achieved by the variance-based measure at a computational cost that is several orders of magnitude lower than when employing the nearest-neighbor-based nonconformity measure.

Keywords: Conformal prediction · Nonconformity measures · Regression · Random forests

1 Introduction

When employing the conformal prediction (CP) framework [13], the probability of making incorrect predictions is bounded by a user-provided confidence threshold. Rather than just providing a single bound on the prediction error for examples drawn from the underlying distribution, CP allows for providing different bounds for different instances, something which may be valuable in many

© Springer International Publishing Switzerland 2016
A. Gammerman et al. (Eds.): COPA 2016, LNAI 9653, pp. 75–89, 2016.
DOI: 10.1007/978-3-319-33395-3_6

different scenarios. For example, in the medical domain, the ability to assess the confidence in predictions related to individual patients, rather than at the group level, may be crucial input for decisions concerning alternative treatments for a specific patient.

CP relies on real-valued functions, called nonconformity measures, that provide estimates for how different a new example is from a set of old examples. It is possible to design many different nonconformity functions for a specific predictive model, and each will result in a different conformal predictor. All such conformal predictors will be *valid* i.e., the probability of excluding the correct label will be less than one minus the confidence level. However, there may be significant differences in terms of *efficiency*, i.e., the sizes of output prediction regions, meaning that the informativeness of the output of different conformal predictors may vary substantially. For classification, efficiency is often measured as the (average) number of labels present in the prediction sets, while for regression, which is the focus of this paper, efficiency is most commonly measured as the (average) size of the intervals.

CP was originally introduced as a transductive approach [5], which requires the learning of a new model for each new test instance to be predicted. Since this in many cases may be computationally prohibitive, inductive conformal prediction (ICP) was suggested [13]. In ICP, which is the focus of this study, a single model is learned from the training data and that model is then used for predicting all test instances. In ICP, however, the calculation of the nonconformity scores (normally) requires comparing predicted values with true target values not seen during training, and the standard procedure to achieve this is to set aside a separate subset of the training examples, called the *calibration set*. However, when the underlying model is an ensemble constructed using bagging, such as a random forest [3], there is also an option to use out-of-bag estimates for the calibration, effectively allowing all training data to be used for constructing the underlying model, something which has been exploited in the context of ICP for bagged ANNs [7] and random forests [6].

Until recently, most studies on ICP conformal regression have focused on one specific underlying model, using a limited number of data sets, making them serve mainly as proofs-of-concept rather than allowing for drawing statistically valid conclusions; see e.g., [8,10]. The apparent need for larger studies evaluating techniques for producing efficient conformal predictors, motivated the study in [6], in which the use of a random forest as the underlying model was compared to existing state-of-the-art conformal regressors, based on neural networks [9] and k-nearest neighbors [11]. A number of nonconformity measures were investigated, including the option to use out-of-bag estimates for the necessary calibration. The results in [6] showed that for almost all confidence levels and using both standard and normalized nonconformity functions, a random forest conformal predictor calibrated using a normalized nonconformity function based on out-of-bag errors of neighboring instances, produced significantly more efficient conformal predictors than the existing alternatives.

However, the use of a nonconformity measure based on the k nearest neighbors requires that access has to be provided to all training instances even at the time the model is deployed, something which occasionally may limit the usefulness of the approach, e.g., when there are size constraints, such as on mobile devices, or when data is highly sensitive and may not be re-distributed. An even more important constraint may be the computation time, both for training and testing. The computational cost of calculating the average error of the k nearest neighbors for each example in the training set is quadratic in the number of examples, hence incurring a substantial additional cost for employing the conformal framework, which may be a limiting factor in particular when handling large training sets. Even for testing, there is an additional cost when using the nearest neighbor nonconformity measure, since the distance of each test instance to all training instances needs to be calculated. To increase the applicability of conformal regression using random forests, there is hence a need for nonconformity measures with lower computational cost. One such candidate approach is to estimate the difficulty of an instance, not by averaging the errors of its neighbors, but by utilizing the fact that each prediction of a random forest is formed by averaging votes of the individual trees in the forest. For difficult cases, one would expect a larger degree of disagreement among the trees, i.e., a higher variance among the individual predictions, than for easier cases. In other words, the variance could be used as an estimate of the difficulty. In fact, this idea is not entirely novel, but was already investigated for k-nearest neighbor regressors in [11], where the variance of the target value of the k neighbors was one of several proposed estimates of difficulty. The main question of this study is whether or not this is an effective approach for forests of regression trees.

In the next section, we formalize the conformal regression framework. In Sect. 3, we describe the current state-of-the-art approach for conformal regression, i.e., random forests using out-of-bag errors of neighboring instances, as well as the proposed approach, which instead of employing the nearest neighbor procedure uses the variance of the predictions made by the individual trees to normalize the prediction regions. The setup for, and the results from, the empirical investigation are presented in Sect. 4. Finally, we summarize the main conclusions from the study and outline directions for future work in Sect. 5.

2 Background

Predictions of a conformal regressor take the form of real-valued intervals (a, b), where $P(a \leq y \leq b) \geq 1 - \delta$ for a test pattern x with true output value y and a user-specified significance level δ. To produce such *prediction intervals*, a conformal regressor utilizes a *nonconformity measure*, which is a real-valued function that measures the strangeness of an example (x, y). This nonconformity measure is typically based on the prediction error of a traditional machine learning model, called the *underlying model* of the conformal regressor. Based on the nonconformity scores of examples with known labels, a conformal predictor uses hypothesis testing to reject (or fail to reject) tentative output values $\tilde{y} \in \mathbb{R}$ at significance δ. For regression problems, the nonconformity measure is most often simply the absolute prediction error [9–11],

$$\alpha_i = A(\boldsymbol{x}_i, y_i, h) = |y_i - \hat{y}_i| = |y_i - h(\boldsymbol{x}_i)|, \tag{1}$$

where h is the underlying model trained on the problem in question, e.g., a regression tree, a neural network or an ensemble model.

To train an inductive conformal regressor, the following procedure is normally used:

1. Divide the training set $Z = \{(\boldsymbol{x}_1, y_1), ..., (\boldsymbol{x}_l, y_l)\}$ into two disjoint subsets Z^t (a proper training set) and Z^c (a calibration set):
   - $Z^t = \{(\boldsymbol{x}_1, y_1), ..., (\boldsymbol{x}_m, y_m)\}$
   - $Z^c = \{(\boldsymbol{x}_{m+1}, y_{m+1}), ..., (\boldsymbol{x}_l, y_l)\}$
2. Train the underlying model h using Z^t.
3. Use the nonconformity measure, e.g. (1), to measure the nonconformity of the examples in Z^c, obtaining a list, sorted in descending order, of calibration scores $S = \alpha_1, ..., \alpha_q$ where $q = |Z^c|$.

When a new test instance $\boldsymbol{x}_j$ arrives, a prediction region is constructed as follows:

1. Obtain a prediction $\hat{y}_j = h(\boldsymbol{x}_j)$.
2. Find the calibration score $\alpha_{s(\delta)}$ where $s(\delta) = \lfloor \delta(q+1) \rfloor$.
3. Using the (partial) inverse of the nonconformity measure, obtain the largest error that is consistent with δ, i.e., $A^{-1}(\alpha_{s(\delta)})$. This is the maximum error made by h on $\boldsymbol{x}_j$ with confidence $1 - \delta$.

If the nonconformity measure in (1) is used, the predictive step simply translates into a prediction region for $\boldsymbol{x}_j$ being constructed as

$$\hat{Y}_j^\delta = \hat{y}_j \pm \alpha_{s(\delta)}, \tag{2}$$

since, with probability $1 - \delta$, the underlying model h will not make an absolute prediction error greater than $\alpha_{s(\delta)}$.

It must be noted that when using (1) and (2), the conformal regressor will, for any specific significance level δ, always produce prediction intervals of the same size for every $\boldsymbol{x}_j$; i.e., the error bounds will not be dependent on properties of a specific test instance. It is, however, possible to introduce *normalized* nonconformity measures, where the absolute error is divided by a term σ_i that is dependent on the prediction instance, usually corresponding to the estimated difficulty of the underlying model for making a correct prediction for that instance; see e.g., [9, 11]:

$$\alpha_i = \frac{|y_i - \hat{y}_i|}{\sigma_i}. \tag{3}$$

With normalized nonconformity measures, the prediction interval for $\boldsymbol{x}_j$ is:

$$\hat{Y}_j^\delta = \hat{y}_j \pm \alpha_{s(\delta)} \sigma_j. \tag{4}$$

The motivation for employing normalized nonconformity functions is that instances estimated to be easier to predict will be assigned narrower intervals than instances that are judged to be more difficult. It should be noted that there are several ways to estimate the difficulty; one suggestion is to train another model for predicting the errors; see e.g., [9]. Other approaches use properties of the underlying model; see e.g., [11].

3 Methods

In this section, we describe the approach for regression conformal prediction using random forests. In particular, we describe three nonconformity measures that will be compared in the empirical investigation: (i) a standard (non-normalized) nonconformity measure, (ii) a nonconformity measure where the difficulty is estimated by the average error of the nearest neighbors, which was shown to result in state-of-the-art performance in [6], and (iii) a variance-based nonconformity measure, originally proposed for k-nearest neighbor classifiers in [11], which previously has not been evaluated for random forests.

3.1 Regression Conformal Prediction using Random Forests

A random forest [3] is an ensemble consisting of *random trees*, which are decision trees generated in a specific way. In order to introduce the necessary diversity, each random tree is trained on a *bootstrap replicate* [2], and only a randomized subset of the attributes are available for the algorithm when optimizing each interior split. The instances that were missing in the bootstrap replicate, for a specific tree, are said to be *out-of-bag* (oob) for that tree. In this study, and similar to [6], we will investigate nonconformity functions that are based on absolute errors, see (1) and (3), where oob instances are used for calculating calibration scores, instead of using a separate calibration set. This means that for each instance in the original training set, only those trees in the generated forest for which the instance is oob, are used for generating the prediction, i.e., instead of $\hat{y}_i = h(x_i)$ in (1) and (3), where h is a random forest, $\hat{y}_i = h_i(x_i)$, where $h_i \subseteq h$. The expected number of trees used to form an oob prediction is approximately 0.368 of the original number of trees, since the probability of including a training example in a bootstrap replicate is about 0.632 [2], leading to that prediction errors on the oob instances can be expected to be at least as large as for independent test instances when using the entire forest, since the underlying model used for the calibration is weaker. Hence, as argued in [6], calculating non-conformity scores using oob instances should lead to valid, although conservative, prediction regions. It should be noted that since all training data can be used for constructing the underlying models, these are typically stronger than the corresponding models trained on a subset, i.e., when excluding the calibration instances, something which was demonstrated in [6] to result in significant efficiency improvements.

3.2 Non-normalized Nonconformity Measure

The first nonconformity measure employs (2), i.e., there is no normalization, so all prediction regions will have identical sizes. It must be noted, however, that out-of-bag instances are used for the calibration instead of a separate calibration set, making it possible to use all available instances for both the training and the calibration. More specifically, when producing the nonconformity score for a calibration instance z_i, the ensemble used for producing the prediction $\hat{y}_i$ consists of all trees that were not trained using z_i, i.e., z_i was out-of-bag for those trees.

3.3   Nearest Neighbor-Based Normalization

The second nonconformity measure employs normalization using (3), i.e., the sizes of the prediction regions vary depending on the estimated difficulty of the instances. Inspired by nonconformity measures proposed for k-nearest neighbor classifiers [11], this nonconformity measure estimates difficulty by the (out-of-bag) error of the k nearest neighbors, with the obvious motivation that low errors for neighboring instances imply a relatively easy part of the feature space. The exact number of neighbors to use is optimized (between 1 and 45) for each training set (more precisely, for each fold, when performing cross-validation), and the k resulting in the smallest average interval size of the resulting conformal regressor is chosen.

The resulting nonconformity measure for an instance $(\boldsymbol{x}_i, y_i)$ is:

$$\alpha_i = \frac{|y_i - \hat{y}_i|}{\mu_i + \beta} \tag{5}$$

where μ_i is an estimate of the difficulty and β is a parameter, used to control the sensitivity of the nonconformity measure. The difficulty estimate for this particular nonconformity measure is the average, distance-weighted, out-of-bag error of the k nearest neighbors:

$$\mu_i = \frac{\sum_{n=1}^{k} o_n/d_n}{\sum_{n=1}^{k} 1/d_n} \tag{6}$$

where $\{o_1, \ldots, o_k\}$ are the out-of-bag errors of the k nearest neighbors and $\{d_1, \ldots, d_k\}$ are the Euclidean distances of the nearest neighbors to $\boldsymbol{x}_i$ plus a small term ϵ (to avoid division by zero).

Using this nonconformity function, the prediction intervals become:

$$\hat{Y}_j^{\delta} = \hat{y}_j \pm \alpha_{s(\delta)}(\mu_j + \beta) \tag{7}$$

When used with random forests and out-of-bag calibration, this nonconformity measure was in [6] shown to outperform all competing approaches, including conformal regressors based on neural networks [9] and k-nearest neighbors [11]. Hence, this particular configuration may be considered as the current state-of-the-art for inductive conformal regression.

3.4   Variance-Based Normalization

The third, and last, nonconformity measure that is evaluated in this study estimates difficulty by the variance of the predictions of the individual trees in the forest. The motivation for this difficulty estimator is that for easier instances, one may expect a higher degree of agreement among the trees in the forest. This nonconformity measure has, again, been studied in the context of k-nearest neighbor classifiers [11], but has not previously been investigated for conformal regressors using random forests. This measure is on the same form as the previous (5), but where μ_i now corresponds to the variance of the individual predictions for an instance $(\boldsymbol{x}_i, y_i)$:

$$\mu_i = \frac{\sum_{n=1}^{s} p_n^2}{s} - \left(\frac{\sum_{n=1}^{s} p_n}{s} \right)^2 \tag{8}$$

where $\{p_1, \ldots, p_s\}$ are the predictions of the trees in the forest for which the instance (x_i, y_i) is out-of-bag.

Using this nonconformity measure, the prediction intervals are, as for the previous measure, calculated using (7).

4 Empirical Evaluation

In this section, we first describe the experimental setup, i.e., what algorithms, datasets and performance metrics have been chosen, and then report the results from the experiment.

4.1 Experimental Setup

For the empirical investigation, all competing methods were re-implemented in the Julia language[1], and a large-scale study, using 33 publicly available data sets from the UCI [1] and Delve [12] repositories, was performed. The considered data sets are small to medium sized; ranging from approximately 500 to 10000 instances. To allow for comparing sizes of prediction regions with the entire output space, the target variable was normalized for each dataset by:

$$\tilde{y}_i = \frac{y_{max} - y_i}{y_{max} - y_{min}} \tag{9}$$

where y_{max} and y_{min} are the highest and lowest output values, respectively, for the dataset. The same normalization was employed also for each input variable, to avoid choice of scale having an impact when calculating Euclidean distances for the nearest neighbor-based nonconformity measure. The latter has neither any effect on the other nonconformity measures nor on the underlying random forest models, i.e., the predictive performance is unaffected.

Regarding parameter values, similar settings as in [6] were employed for all data sets and methods. Specifically, all random forests consisted of 500 random trees, the sensitivity parameter β was set to 0.01 while the parameter ϵ was set to 0.001. A ten-fold cross validation scheme was adopted with all reported values being averaged over the ten folds. Results are reported for three confidence levels: 90 %, 95 % and 99 %.

For each method and dataset in the experiments, the *error rate*, i.e., the fraction of target values in the test set that fall outside the predicted regions, and the *efficiency*, i.e., the size of the predicted intervals, are measured. For valid conformal predictors, the error rate should not (in the long run) be higher than one minus the chosen confidence threshold. Hence, by investigating the error rate, we may confirm (or reject) that a certain conformal predictor actually is valid. Note that this is here considered to be a binary property, i.e., we do not

---

[1] www.julialang.org.

consider one method to be more valid than another. Given that we have a set of valid regression conformal predictors, the perhaps most interesting aspect to compare is the size of the predicted regions, as this directly corresponds to how informative these regions are. Such a comparison could be done in different ways, e.g., comparing extreme values, but we have similar to [6] opted for comparing the average sizes over all prediction regions.

In order to allow for a comparison of the computational cost for generating and applying the different nonconformity measures, i.e., during training and testing, respectively, the CPU times for these activities were recorded, separately from the time taken to build the forests and obtaining predictions from the individual trees. In the experiment, a DELL T7910 with two 14-core 2.6 GHz CPUs (E5-2697v3) with 64 GB RAM was employed. The generation and application of all nonconformity measures was performed on a single core only, while the forest construction and predictions utilized all cores in parallel[2].

To analyze any differences in efficiency between the two normalized approaches, the correlation coefficient between the estimated difficulty of the test instances and the actual prediction error are reported for each method. The expectation is that a higher correlation leads to more efficient predictions.

4.2   Experimental Results

Table 1 shows the error rates, i.e., the fraction of test instances for which the true target value falls outside the predicted region, of three methods utilizing different nonconformity functions: using no normalization (M1); using nearest-neighbor normalization (M2); and, using variance-based normalization (M3). Looking at these results, it is apparent that all three methods behave as expected for valid predictors: the error rates, for each data set, lie close to the predetermined significance level.

A statistical analysis of the error rates at the three confidence levels presented (90 %, 95 % and 99 %), using a Friedman test followed by a Nemenyi post-hoc test (with alpha=0.05) [4], shows that: (i) M3 has a significantly lower error rate than both M1 and M2 for the 90 % level, (ii) M3 has a significantly lower error rate than M1 for the 95 % level, and (iii) M3 has a significantly lower error rate than M2 for the 99 % confidence level. Hence, the variance-based approach clearly seems to be the most conservative of the three methods.

Looking at the interval sizes tabulated in Table 2, while remembering that the output was normalized so that an interval size of 1.0 would cover the entire range of the target values, it can be seen from the averaged values that the best method at the 90 % confidence level returned prediction regions covering, approximately, 21 % of the range. The corresponding average values for the 95 % and 99 % confidence levels are (approximately) 26 % and 38 %, respectively. Clearly, these prediction regions must be considered informative. An analysis of the interval sizes, using the same statistical test as earlier, reveals that there is no significant difference between M2 and M3 for any of the three confidence levels, while both

---

[2] The Julia implementation can be obtained from the first author upon request.

Table 1. Error rates

Confidence dataset \ Technique	0.90			0.95			0.99		
	M1	M2	M3	M1	M2	M3	M1	M2	M3
abalone	.099	.101	.104	.050	.053	.049	.010	.013	.010
anacalt	.099	.082	.094	.047	.036	.047	.008	.012	.009
bank8fh	.100	.099	.098	.049	.050	.047	.011	.011	.009
bank8fm	.099	.098	.093	.049	.049	.048	.010	.010	.009
bank8nh	.100	.101	.098	.050	.051	.050	.010	.011	.010
bank8nm	.100	.102	.098	.050	.051	.048	.009	.011	.010
boston	.107	.101	.099	.049	.042	.036	.008	.010	.010
comp	.096	.100	.098	.049	.050	.050	.010	.011	.010
concreate	.098	.081	.100	.050	.044	.049	.010	.008	.008
cooling	.095	.092	.092	.052	.050	.050	.012	.013	.012
deltaA	.101	.103	.100	.050	.051	.049	.009	.010	.010
deltaE	.099	.103	.099	.051	.053	.048	.010	.012	.010
friedm	.097	.098	.093	.050	.046	.050	.008	.004	.007
heating	.102	.081	.092	.050	.048	.053	.005	.006	.009
istanbul	.105	.108	.099	.050	.052	.050	.007	.011	.007
kin8fh	.099	.098	.099	.050	.049	.049	.010	.009	.009
kin8fm	.099	.094	.094	.049	.043	.047	.010	.007	.009
kin8nh	.099	.100	.098	.049	.048	.048	.009	.009	.008
kin8nm	.096	.092	.096	.049	.047	.047	.010	.009	.008
laser	.098	.088	.090	.047	.041	.049	.009	.009	.007
mg	.097	.097	.095	.046	.055	.051	.009	.013	.012
mortage	.091	.087	.091	.044	.034	.044	.009	.007	.008
plastic	.101	.107	.098	.052	.050	.050	.008	.015	.007
puma8fh	.097	.100	.097	.050	.051	.048	.009	.011	.010
puma8fm	.100	.099	.100	.050	.051	.049	.009	.010	.008
puma8nh	.100	.102	.096	.051	.050	.047	.010	.010	.009
puma8nm	.095	.096	.095	.048	.049	.046	.009	.011	.009
quakes	.100	.107	.096	.051	.060	.053	.014	.026	.019
stock	.094	.088	.099	.046	.040	.046	.008	.003	.009
treasury	.099	.095	.103	.048	.042	.045	.011	.012	.010
wineRed	.101	.104	.098	.051	.054	.048	.009	.014	.010
wineWhite	.103	.107	.101	.048	.053	.047	.011	.011	.008
wizmir	.095	.106	.089	.047	.046	.045	.010	.012	.012
Mean	.099	.097	.097	.049	.048	.048	.009	.011	.009
Mean rank	2.26	2.21	1.53	2.32	2.03	1.65	1.95	2.39	1.65

Table 2. Region sizes.

Confidence dataset \ Technique	0.90			0.95			0.99		
	M1	M2	M3	M1	M2	M3	M1	M2	M3
abalone	.234	.214	.214	.321	.274	.282	.544	.463	.495
anacalt	.139	.081	.107	.258	.092	.126	.501	.190	.221
bank8fh	.300	.290	.268	.377	.361	.342	.533	.538	.585
bank8fm	.139	.131	.123	.175	.158	.145	.251	.211	.191
bank8nh	.322	.307	.281	.447	.420	.414	.789	.744	.782
bank8nm	.145	.121	.111	.210	.160	.141	.399	.245	.217
boston	.193	.192	.200	.276	.254	.253	.605	.432	.418
comp	.086	.077	.083	.114	.098	.107	.187	.153	.170
concreate	.204	.208	.184	.258	.270	.235	.475	.473	.362
cooling	.170	.107	.150	.216	.124	.184	.287	.146	.243
deltaA	.117	.108	.113	.154	.139	.141	.260	.212	.228
deltaE	.174	.170	.172	.215	.215	.214	.315	.305	.304
friedm	.215	.205	.217	.258	.243	.269	.360	.319	.406
heating	.070	.058	.065	.087	.068	.078	.168	.094	.102
istanbul	.260	.247	.257	.318	.315	.336	.491	.497	.494
kin8fh	.241	.240	.240	.291	.285	.285	.398	.372	.375
kin8fm	.134	.123	.132	.166	.144	.160	.245	.183	.218
kin8nh	.413	.404	.408	.488	.472	.478	.622	.595	.613
kin8nm	.331	.303	.321	.396	.350	.374	.527	.445	.478
laser	.044	.039	.041	.085	.054	.059	.330	.150	.141
mg	.243	.172	.163	.341	.221	.201	.596	.322	.336
mortage	.022	.019	.021	.036	.027	.032	.073	.044	.059
plastic	.549	.545	.592	.644	.637	.734	.807	.851	.943
puma8fh	.470	.446	.444	.565	.532	.529	.741	.724	.754
puma8fm	.210	.204	.201	.254	.243	.240	.341	.323	.322
puma8nh	.438	.427	.416	.543	.518	.503	.731	.697	.697
puma8nm	.202	.199	.201	.243	.238	.233	.345	.328	.310
quakes	.556	.540	.605	.705	.681	.751	1.000	.900	.942
stock	.076	.074	.074	.093	.089	.088	.158	.131	.124
treasury	.026	.022	.025	.042	.030	.039	.088	.051	.071
wineRed	.366	.375	.336	.495	.499	.452	.734	.721	.636
wineWhite	.321	.320	.289	.416	.420	.372	.644	.662	.551
wizmir	.059	.058	.059	.074	.072	.073	.139	.126	.125
Mean	.226	.213	.216	.290	.264	.269	.445	.383	.391
Mean rank	2.79	1.42	1.79	2.76	1.61	1.64	2.73	1.55	1.73

M2 and M3 result in significantly smaller interval sizes than M1 for all three confidence levels (with p-values much smaller than 0.01).

Table 3 displays execution times for the three different methods tested. First listed is the total time (in seconds) for the tasks common to all methods of training the underlying model (random forest), collecting out-of-bag predictions and obtaining the individual predictions for the test instances. As expected, only small variations are observed, since these tasks are identical for all three approaches.

Second, the total number of seconds required to generate the nonconformity measure using the out-of-bag instances is listed. Here, there is a clear difference between the three methods. M1 requires only that the errors on the out-of-bag instances are computed and ordered, which is a fairly quick operation. M2, on the other hand, requires an extra (particularly costly) step of making, for each out-of-bag instance, an additional prediction using the nearest-neighbor procedure to calculate the normalization term of the nonconformity measure. Finally, M3, for which normalization does not require any additional predictive step, the calculation of nonconformity scores comes with very little overhead compared to the non-normalized variant M1.

Listed in the third column is the total time (in seconds) required to calculate prediction regions for the test set. Again, the time required for making predictions using the variance-based M3 is only marginally longer than for the non-normalized M1, while M2 again incurs a very large overhead.

It should be noted that the observed execution times are dependent on the particular implementation of the algorithms, and possibly some of the performance differences could be reduced by carefully optimizing the code. However, there is an inherent difference in computational complexity of the underlying algorithms, which will not disappear even with smarter implementations. Comparing the computational cost that is specific to performing conformal prediction, i.e., not including the time for building and obtaining predictions from the underlying model, the variance-based approach is in this experiment several orders of magnitude faster than the nearest neighbor approach (the former is on average over twenty thousand times faster than the latter) and this gap will most likely remain wide even with a substantially more efficient implementation of the k-nearest neighbor procedure.

Finally, in order to investigate how well the difficulty estimates employed by the nearest-neighbor and the variance-based approaches actually work, we investigated the correlation coefficients between $\mu_i + \beta$ and the test error for the two normalized approaches. The results are displayed in Table 4. When testing for significant differences, the p-value is 0.056 in favor of M3 over M2, hence indicating that variance in fact may be a more effective way of ordering instances according to expected test error than employing the nearest-neighbor procedure. This difference obviously does not directly carry over to a corresponding difference in region size, as the latter was found above to be insignificant (Table 2). However, the importance of correctly ranking the instances according to difficulty is demonstrated by the fact that the method with the highest correlation

Table 3. Time taken (in seconds) to build and obtain predictions from the underlying models (identical tasks for all methods), to generate the nonconformity functions and to calculate prediction regions for the test set

Dataset \ Technique	Common tasks			Calibration			Application		
	M1	M2	M3	M1	M2	M3	M1	M2	M3
abalone	2.91	3.02	2.98	.002	35.1	.003	.000	4.00	.000
anacalt	1.02	.98	1.03	.002	31.8	.003	.000	3.62	.000
bank8fh	6.54	6.60	6.74	.003	151.8	.005	.000	17.19	.000
bank8fm	6.67	6.67	6.78	.005	150.3	.005	.000	17.11	.000
bank8nh	6.47	6.54	6.49	.005	151.0	.006	.000	17.24	.000
bank8nm	6.36	6.43	6.53	.003	150.9	.005	.000	17.21	.000
boston	.34	.33	.33	.000	.4	.000	.000	.05	.000
comp	6.67	6.83	6.90	.003	150.3	.007	.000	17.17	.000
concreate	.63	.64	.62	.000	1.7	.001	.000	.19	.000
cooling	.26	.27	.27	.000	1.0	.001	.000	.11	.000
deltaA	5.35	5.38	5.31	.011	108.5	.005	.000	12.40	.000
deltaE	7.15	7.36	7.37	.004	207.1	.007	.000	23.78	.001
friedm	.77	.78	.77	.000	2.3	.001	.000	.27	.000
heating	.27	.28	.27	.000	.9	.001	.000	.10	.000
istanbul	.36	.36	.36	.000	.5	.000	.000	.05	.000
kin8fh	5.93	5.96	6.19	.003	148.0	.007	.000	16.91	.000
kin8fm	5.94	6.06	6.09	.003	147.6	.005	.000	16.85	.001
kin8nh	6.21	6.26	6.29	.003	147.3	.006	.000	16.89	.000
kin8nm	6.07	6.10	6.18	.003	149.2	.007	.000	17.07	.000
laser	.64	.63	.63	.000	1.7	.001	.000	.19	.000
mg	.93	.94	.95	.001	3.2	.001	.000	.36	.000
mortage	.74	.73	.74	.000	1.8	.001	.000	.20	.000
plastic	.49	.50	.49	.001	4.7	.001	.000	.53	.000
puma8fh	6.24	6.31	6.33	.004	149.6	.006	.000	17.04	.000
puma8fm	6.51	6.26	6.26	.003	150.0	.005	.000	17.14	.000
puma8nh	6.22	6.33	6.21	.005	148.7	.006	.000	17.00	.000
puma8nm	6.11	6.22	6.19	.003	146.2	.005	.000	16.71	.000
quakes	1.53	1.48	1.49	.001	8.3	.001	.000	.94	.000
stock	.62	.61	.64	.000	1.5	.001	.000	.17	.000
treasury	.71	.71	.72	.000	1.9	.001	.000	.23	.000
wineRed	.90	.90	.93	.001	4.5	.002	.000	.51	.000
wineWhite	3.32	3.24	3.25	.002	50.6	.017	.000	5.82	.000
wizmir	1.05	1.06	1.06	.001	3.6	.001	.000	.40	.000
Mean	3.39	3.42	3.44	.002	73.1	.004	.000	8.35	.000
Mean rank	1.73	2.03	2.24	1.03	3.00	1.97	1.00	3.00	2.00

Table 4. Correlation between difficulty and test error

Dataset \ Technique	Correlation	
	M2	M3
abalone	.360	.372
anacalt	.853	.825
bank8fh	.172	.300
bank8fm	.404	.498
bank8nh	.196	.272
bank8nm	.602	.670
boston	.361	.429
comp	.443	.346
concreate	.352	.450
cooling	.821	.649
deltaA	.414	.425
deltaE	.176	.204
friedm	.348	.040
heating	.701	.628
istanbul	.072	.129
kin8fh	.221	.226
kin8fm	.557	.272
kin8nh	.248	.224
kin8nm	.468	.317
laser	.571	.695
mg	.668	.764
mortage	.667	.652
plastic	-.082	-.082
puma8fh	.264	.298
puma8fm	.234	.274
puma8nh	.241	.345
puma8nm	.222	.256
quakes	.133	.156
stock	.397	.348
treasury	.713	.594
wineRed	.238	.423
wineWhite	.257	.435
wizmir	.193	.214
Mean	.378	.383
Mean rank	1.67	1.33

coefficient of the two for each dataset, also produces the smallest average prediction region for 21 out of 33 cases. The probability of observing this (or a larger) number is only 0.081 if the resulting region size would be independent of this correlation.

5   Concluding Remarks

In this paper, we have presented a large-scale empirical evaluation of conformal regression approaches using random forests with out-of-bag calibration. We have compared a variance-based nonconformity measure, which has previously not been evaluated in this context, to a standard (non-normalized) nonconformity measure as well as to one measure based on k-nearest neighbors, which previously was found to give state-of-the-art performance. The experimental results in this study show that both the nearest-neighbor-based and the variance-based measures significantly outperform the non-normalized measure, while no significant difference in efficiency between the two normalized approaches is observed. Moreover, the evaluation shows that state-of-the-art performance is achieved by the variance-based measure at a computational cost that is several orders of magnitude lower than when employing the nearest-neighbor-based nonconformity measure.

There are several possible directions for future research. One direction concerns refining the rather straightforward difficulty estimate further, e.g., by not only considering variance of the ensemble member predictions, but also estimates of uncertainty for the individual predictions. Other directions for future research include investigating ways of combining several different difficulty estimates and evaluating the alternative nonconformity measures for other ensemble approaches for which out-of-bag estimates can be obtained.

Acknowledgments. This work was supported by the Swedish Foundation for Strategic Research through the project High-Performance Data Mining for Drug Effect Detection (IIS11-0053), the Vinnova program for Strategic Vehicle Research and Innovation (FFI)-Transport Efficiency, and the Knowledge Foundation through the project Data Analytics for Research and Development (20150185).

References

1. Bache, K., Lichman, M.: UCI machine learning repository (2013). http://archive. ics.uci.edu/ml
2. Breiman, L.: Bagging predictors. Mach. Learn. **24**(2), 123–140 (1996)
3. Breiman, L.: Random forests. Mach. Learn. **45**(1), 5–32 (2001)
4. Demšar, J.: Statistical comparisons of classifiers over multiple data sets. J. Mach. Learn. Res. **7**, 1–30 (2006)
5. Gammerman, A., Vovk, V., Vapnik, V.: Learning by transduction. In: Proceedings of the Fourteenth Conference on Uncertainty in Artificial Intelligence, pp. 148–155. Morgan Kaufmann (1998)

6. Johansson, U., Boström, H., Löfström, T., Linusson, H.: Regression conformal prediction with random forests. Mach. Learn. **97**(1–2), 155–176 (2014)
7. Löfström, T., Johansson, U., Boström, H.: Effective utilization of data in inductive conformal prediction. In: The 2013 International Joint Conference on Neural Networks (IJCNN). IEEE (2013)
8. Papadopoulos, H.: Inductive conformal prediction: theory and application to neural networks. Tools Artif. Intell. **18**(2), 315–330 (2008)
9. Papadopoulos, H., Haralambous, H.: Reliable prediction intervals with regression neural networks. Neural Netw. **24**(8), 842–851 (2011)
10. Papadopoulos, H., Proedrou, K., Vovk, V., Gammerman, A.J.: Inductive confidence machines for regression. In: Elomaa, T., Mannila, H., Toivonen, H. (eds.) ECML 2002. LNCS (LNAI), vol. 2430, pp. 345–356. Springer, Heidelberg (2002)
11. Papadopoulos, H., Vovk, V., Gammerman, A.: Regression conformal prediction with nearest neighbours. J. Artif. Intell. Res. **40**(1), 815–840 (2011)
12. Rasmussen, C.E., Neal, R.M., Hinton, G., van Camp, D., Revow, M., Ghahramani, Z., Kustra, R., Tibshirani, R.: Delve data for evaluating learning in valid experiments (1996). www.cs.toronto.edu/delve
13. Vovk, V., Gammerman, A., Shafer, G.: Algorithmic Learning in a Random World. Springer, Heidelberg (2006)

Binary Relevance Multi-label Conformal Predictor

Antonis Lambrou[1(✉)] and Harris Papadopoulos[1,2]

[1] Computer Learning Research Centre Royal Holloway,
University of London, London, UK
{A.Lambrou,H.Papadopoulos}@cs.rhul.ac.uk
[2] Computer Science and Engineering Department,
Frederick University, Nicosia, Cyprus
H.Papadopoulos@frederick.ac.cy

Abstract. The Conformal Prediction (CP) framework can be used for obtaining reliable confidence measures in Machine Learning applications. The confidence measures are guaranteed to be valid under the assumption that the data used are identically and independently distributed (i.i.d.). In this work, we extend the CP framework for multi-label classification, where an instance can belong to multiple classes in parallel. Applications include image tagging, document classification, and music classification. We give an overview of the Conformal Prediction framework, and we describe the developed Binary Relevance Multi-Label Conformal Predictor (BR-MLCP). We propose a new measure of confidence using Chebyshev's inequality together with the hamming loss metric. Our experimental results demonstrate the reliability of our new confidence measure.

Keywords: Multi-label · Conformal prediction · Confidence measures

1 Introduction

Conformal Predictors (CPs) [1] are algorithms that can provide predictions complemented with reliable confidence measures, which are guaranteed to be valid under the assumption that the data used are identically and independently distributed (i.i.d.). The CP framework was first proposed in [2] and a more recent description can be found in [3]. CPs are built using classical machine learning algorithms, called underlying algorithms, and complement the predictions of the underlying algorithms with measures of confidence. Many CPs have been built to date, based on various algorithms such as Support Vector Machines [4], k-Nearest Neighbours [5], and Random Forests [6]. The CP framework has been successfully applied to medical diagnostic problems, such as ovarian cancer diagnosis [7], breast cancer diagnosis [8], and acute abdominal pain diagnosis [9,10]. Other extensions of CPs include information fusion [11], and feature selection [12].

The CP framework can be extended to multi-label classification, where a data instance can be associated with multiple classes in parallel and the predictions of such data may accommodate more than one labels. Applications include

© Springer International Publishing Switzerland 2016
A. Gammerman et al. (Eds.): COPA 2016, LNAI 9653, pp. 90–104, 2016.
DOI: 10.1007/978-3-319-33395-3_7

image tagging, document classification, gene function categorization, and music classification. For example, in document classification, a specific document which contains both religious and political issues can have both labels: one label for class "politics", and one for class "religion". Multi-label algorithms are generally categorized into two groups based on the transformation method that is used. One group is using Pattern Transformation (PT), where the multi-labelled data are split into several single labelled data, and then traditional machine learning algorithms can be applied for classification. The second group is using Algorithmic Adaptation (AA), where the underlying algorithm is transformed in order to construct a mutli-label classifier. An overview of multi-label classification is provided in [13]. In a related study, a CP was developed for multi-label classification using powersets [15]. The powerset method (which falls into the PT group) transforms the multi-label classification task into single label classification by mapping each combination of the available labels into single labelled classes. Another study, which follows another PT approach, can be found in [16,17], where the multi-labelled data are decomposed into multiple binary labelled datasets (Binary Relevance approach), and a CP is applied on each subset.

In this paper, we extend the work in [17], and we propose a confidence measure using the hamming loss metric, which is the most common evaluation measure in the setting of multi-label classification. Our proposed confidence measure allows us to produce multi-label prediction regions with at most ε chance of having a hamming loss more than some threshold h. In other words, we can guarantee under i.i.d. assumption, that hamming loss in our multi-label predictions will not exceed h given some confidence $1 - \varepsilon$. In the next sections, we give an overview of the CP framework, we describe the developed Binary Relevance Multi-Label Conformal Predictor (BR-MLCP), and we provide an upper bound of hamming loss using the CP framework and Chebychev's inequality. Finally, we provide experimental results that demonstrate the reliability of our confidence measure.

2   Conformal Prediction Framework

Provided a training dataset, CPs output predictions for new instances together with valid confidence measures, based on the assumption that the given data are identically and independently distributed (i.i.d.). CPs generate prediction regions (sets of possible labels for a new instance), such that the error rate of the prediction regions is guaranteed to not exceed a given significance level in the long run. Additionally, CPs can be configured to output single predictions (instead of prediction regions), together with valid confidence measures. We explain how this is done in the following paragraphs.

A training set is of the form $\{(x_1, y_1), \ldots, (x_n, y_n)\}$, where x_i is a vector of real-valued attributes and $y_i \in \{Y_1, Y_2, \ldots, Y_c\}$ is a label given to the instance x_i. Given a new instance x_{n+1} with unknown label, the target is to find the likelihood of correctness for each possible label $Y_g \in \{Y_1, Y_2, \ldots, Y_c\}$ that can be given to x_{n+1}. A set of steps are performed for each assumed label, in order to obtain the likelihood:

1. The new instance x_{n+1} is appended in the training set together with the assumed label Y_g.
2. An underlying machine learning algorithm is trained on the extended training set

$$\{(x_1, y_1), \ldots, (x_{n+1}, Y_g)\}. \tag{1}$$

3. The underlying algorithm is transformed in order to generate a non-conformity score for each of the instances in (1). A non-conformity score indicates how different (or strange) an instance x_i is for its given label y_i, compared to the other instances in (1).
4. The following p-value function is used to calculate how likely the assumed label is of being correct:

$$p(Y_g) = \frac{\#\{i = 1, \ldots, n+1 : a_i \geq a_{n+1}\}}{n+1}, \tag{2}$$

which compares the non-conformity score a_{n+1} of (x_{n+1}, Y_g) with all the other non-conformity scores of the rest of the instances in the extended training set.

Given the true label y_{n+1}, the p-value function in (2), satisfies the following property for all probability distributions and for any significance level ε:

$$P(p(y_{n+1}) \leq \varepsilon) \leq \varepsilon. \tag{3}$$

In fact, the p-value function is a test function which measures how likely the dataset is of being i.i.d. If the p-value is lower than a given ε, it is because we either have non i.i.d. data, or because some event has happened with probability less than or equal to ε. Based on the assumption that the data are i.i.d., we realise that if we include in our predictions all assumed labels that are assigned a p-value greater than a given significance level ε, then the probability of missing the true label of an instance will be less than or equal ε. In the case that all p-values are less than ε, the label with the highest p-value is included to ensure that the prediction regions will always contain at least one prediction. This step does not increase the probability of error. The definition of a prediction region is given as

$$R = \{Y_g : p(Y_g) > \varepsilon\} \cup \left\{arg \max_{g=1,\ldots,c} (p(Y_g))\right\}. \tag{4}$$

In the long run, these regions will make errors at a rate of at most ε. Therefore, the confidence is calculated as $1 - \varepsilon$. The formal definition of the Conformal Predictor algorithm is given in Algorithm 1.

By preference, we may output only single labels (forced predictions) instead of prediction regions. In forced prediction, only the label with the highest p-value is given as a prediction, together with a confidence measure which is 1 minus the second largest p-value. The confidence measure indicates how likely the prediction is of being correct, with respect to the rest of the possible labels.

Algorithm 1. Conformal Predictor

Input: training set $\{(x_1, y_1), \ldots, (x_n, y_n)\}$, new instance $= x_{n+1}$, possible labels
    $Y_g \in \{Y_1, Y_2, \ldots, Y_c\}$, significance level ε
for $g = 1$ **to** c **do**
    Train the underlying algorithm on the extended set
    $\{(x_1, y_1), \ldots, (x_n, y_n), (x_{n+1}, Y_g)\}$;
    Supply the input patterns $x_1, \ldots, x_{n+1}$ to the underlying algorithm to
    obtain the respective non-conformity scores $a_1, \ldots, a_{n+1}$;
    Calculate the p-value $p(Y_g) = \frac{\#\{i=1,\ldots,l+1:a_i \geq a_{n+1}\}}{n+1}$;
end
Output:
    Prediction Region $R = \{Y_g : p(Y_g) > \varepsilon\} \cup \{arg\max_{g=1,\ldots,c}(p(Y_g))\}$

2.1 Non-conformity

The k-Nearest Neighbours (k-NN) method computes the distance of a test instance from the other instances that are provided in the training set, and finds its k nearest instances. The prediction of the algorithm is the class of the majority of the k instances. In the case of building a CP based on k-NN (k-NN-CP), we use the distances of the k nearest instances to define a non-conformity measure. The simplest approach is to calculate the total of distances of the k instances that belong to the class of instance x_i, since the nearer the instance is to its class, the less strange it is. Nonetheless, for a more efficient non-conformity measure we also take into consideration the distances of the k nearest instances that belong to other classes, since the nearer the instance x_i is to the other classes the more strange it is. We build a k-NN-CP using the non-conformity measure defined in [18]:

$$a_i = \frac{\sum_{j=1,\ldots,k} s_{ij}}{\sum_{j=1,\ldots,k} o_{ij}}, \tag{5}$$

where s_{ij} is the jth shortest distance of x_i from the instances of the same class, and o_{ij} is the jth shortest distance of x_i from the instances of other classes.

3 Developed Algorithm

In multi-label classification, a training set of the form $D = \{(x_1, \psi_1), \ldots, (x_n, \psi_n)\}$ is given, where x_i is an input vector of real-valued attributes, and the instances can be labelled as $\psi_i \subseteq \{Y^1 \times Y^2 \times \ldots \times Y^c\}$, where each $Y^g \in \{y_g^1, y_g^0\}$. Instances that belong to class Y^g are labelled y_g^1, and y_g^0 otherwise. One possible approach to solve a multi-label problem is to decompose it into c single-label binary classification problems (Binary Relevance approach in [17]). The original dataset D is copied into datasets $D_1, \ldots, D_c$, and for each D_g we label as y_g^1 the instances that originally have label y_g^1 in the multi-label ψ_i, and y_g^0 otherwise.

We use a CP on each dataset D_g separately, and given a new instance x_{n+1} and a desirable significance level ε_g, each CP provides a prediction region r_g for class Y^g (as in usual single-label classification). The prediction region r_g states whether the new instance belongs to class Y^g or not, or whether there is uncertainty at the given significance level. We then combine all r_g to provide the prediction region for the multi-label classification task:

$$R = r_1 \times \ldots \times r_c. \tag{6}$$

As shown in property (3), the probability of each r_g missing the true binary label for class Y^g, given a significance level ε_g, is at most ε_g. According to the Bonferroni general inequality, which can be applied for multiple tests, we may state that the probability of the true multi-label ψ_{n+1} not being in R is at most the sum of the upper bound probabilities of the individual r_g sets missing the true binary label:

$$P(\psi_{n+1} \notin R) \leq \sum_{g=1}^{c} \varepsilon_g. \tag{7}$$

Therefore, we have multi-label prediction regions, for which the error rate is

$$\varepsilon \leq \sum_{g=1}^{c} \varepsilon_g. \tag{8}$$

Therefore, for a confidence level $1 - \varepsilon$ in R we set the significance level for each $r_g, g = 1, \ldots, c$ to

$$\varepsilon_g = \frac{\varepsilon}{c}. \tag{9}$$

Alternatively, we may set each ε_g to the second largest p-value provided by each CP, such that each r_g contains a single prediction for the new instance. Thus, the final prediction region R will also contain a single multi-label, which can be considered as a forced prediction for the new instance. The prediction of the multi-label can be complemented with confidence measure $1 - \varepsilon$. The algorithm of the Binary Relevance Multi-Label Conformal Predictor (BR-MLCP) is given in [17] and in Algorithm 2.

3.1 Prediction Regions Based on Hamming Loss

In inequality (8), we consider the error rate with respect to each multi-label prediction as a whole prediction. If the prediction contains even a single binary miss-classification, then the whole multi-label prediction is considered wrong. A more common evaluation metric used for multi-label prediction is the hamming loss metric. Given two sets a and b their hamming loss is calculated as

$$H(a, b) = \#\{g : a_g \neq b_g\}. \tag{10}$$

Given the true multi-label ψ_i of an instance x_i, and a prediction region R_i, the hamming loss of R_i is defined as

$$HL(\psi_i, R_i) = \min_{\pi \in R_i} H(\psi_i, \pi), \tag{11}$$

Algorithm 2. Binary Relevance Multi-Label Conformal Predictor (BR-MLCP[17])

Input: training set $D = \{(x_1, \psi_1), ..., (x_n, \psi_n)\}$, new instance x_{n+1}, possible
      labels $\{Y^1, Y^2, ..., Y^c\}$, significance level ε
for $g = 1$ **to** c **do**
    for $b = 0$ **to** 1 **do**
        $D_g = \{(x_1, Y_1^g), ..., (x_n, Y_n^g), (x_{n+1}, y_g^b)\}$;
        Train the underlying algorithm on the extended set D_g;
        Supply the input patterns $x_1, ..., x_{n+1}$ to the underlying algorithm to
        obtain the respective non- conformity scores $a_1, ..., a_{n+1}$;
        Calculate the p-value $p(y_g^b) = \frac{\#\{i=1,...,l+1:a_i \geq a_{n+1}\}}{n+1}$;
    end
    $r_g = \{y_g^b : p(y_g^b) > \varepsilon/c\} \cup \{arg\max_{g=1,...,c}(p(y_g^b))\}$;
end
Output:
    Prediction Region $R = r_1 \times ... \times r_c$.

---

where π is a multi-label prediction contained in prediction region R_i. We can state that an error occurs only when the hamming loss of a prediction region is above a pre-defined value. Let us denote for $g = 1, ..., c$, $e_g = 1$ if there is a loss in the prediction of class Y^g, and $e_g = 0$ otherwise. By setting the significance level of the gth CP to ε_g for $g = 1, ..., c$, we have

$$P(e_1 = 1) \leq \varepsilon_1; ...; P(e_c = 1) \leq \varepsilon_c. \tag{12}$$

If we allow a hamming loss level h, then the overall prediction is wrong when $e_1 + ... + e_c \geq h + 1$ by definition. As a result of (12), the expected value of $e_1 + ... + e_c$ is at most $\varepsilon_1 + ... + \varepsilon_c$. Consequently, by Chebyshev's inequality we get:

$$P(e_1 + ... + e_c \geq h + 1) \leq \frac{\varepsilon_1 + ... + \varepsilon_c}{h + 1}. \tag{13}$$

In order to show that the upper bound in (13) is optimal, let us assume the case where $\varepsilon_1 = \cdots = \varepsilon_c$, and

$$P\left(\sum_{g=1}^{c} e_g = m\right) = 0, \tag{14}$$

for $m > 0$ and $m \neq h + 1$. This means that the probability of each possible combination of exactly $h + 1$ losses becomes

$$\frac{\varepsilon_g}{C_{c-1}^h}, \tag{15}$$

since each e_g has at most probability ε_g and this is divided between the C_{c-1}^h possible combinations of other losses, which together with e_g result in exactly

$h + 1$ losses. There are C_c^{h+1} possible combinations that give exactly $h + 1$ losses, therefore the total probability of having a hamming loss of more than h is

$$\frac{\varepsilon_g}{(C_{c-1}^h)} \cdot C_c^{h+1} = \frac{c\varepsilon_g}{(h+1)}. \tag{16}$$

This is equal to (13) when $\varepsilon_1 = \cdots = \varepsilon_c$.

As a result of (13) in order to produce multi-label prediction regions with at most ε chance of having a hamming loss more than h, the significance level of the gth CP for $q = 1, \ldots, c$ could be set to

$$\varepsilon_g = \frac{\varepsilon(h+1)}{c}. \tag{17}$$

Alternatively, we can consider the percentage of hamming loss as an error measure. Let us define hamming loss HP as the percentage of errors amongst all predicted labels. By property (3), the probability of each loss in $H(\psi_i, R_i)$ is at most ε_g, and using equation (7), the percentage of hamming loss is

$$HP \leq \sum_{g=1}^{c} \frac{\varepsilon_g}{c}. \tag{18}$$

Therefore, we may provide prediction regions such that the percentage of hamming loss will be at most HP, at a confidence level $1 - HP$. This measure of confidence can also be found in [16].

4   Experiments on Multi-label Datasets

In this section, we evaluate the BR-MLCP and the proposed confidence measure. In our evaluation process, we copy the original datasets into binary class datasets as explained in Sect. 3, and for each subset we apply the Correlation-Based Feature Selection (CBFS) method in order to reduce the number of features. We then apply 10-fold cross validation on each of the reduced datasets. The folds are identical for all datasets. Each test instance on each dataset is given a possible label (y_g^1 or y_g^0), and the test instance is added to the training set. The underlying algorithm is trained on the extended training-set and provides non-conformity scores. A p-value is then generated for each possible binary label given to the test instance. Once we have p-values from all CPs, we apply equation (6) to provide a prediction region for the test instance, given a pre-defined confidence level, or a forced prediction.

4.1   Music into Emotions

We experiment on a multi-label dataset for classifying music into emotions [14]. The Music Emotions dataset contains 593 songs with a total of 72 rhythmic and timbre features in each song. There are 6 possible classes that each song can belong to. The classes and the number of instances in each one are listed in Table 1. As baseline, we provide the average hamming loss of our forced predictions which is 18.77 %. This result is comparable with the results provided in [14], which give an overall hamming loss of 19.43 % for the related Binary Relevance algorithm.

Table 1. Class distribution for the Emotions dataset.

Label	Class	# of instances
1	amazed-surprised	173
2	happy-pleased	166
3	relaxing-calm	264
4	quiet-still	148
5	sad-lonely	168
6	angry-fearful	189

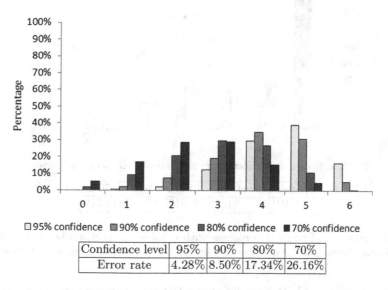

Confidence level	95%	90%	80%	70%
Error rate	4.28%	8.50%	17.34%	26.16%

Fig. 1. Percentages of prediction regions with number of uncertain labels for different levels of confidence, and their respective error rates on the Emotions dataset.

In Fig. 1, we provide the results of the BR-MLCP using (8). The figure shows the distribution of the prediction regions according to the number of uncertain labels, at four different levels of confidence (95 %, 90 %, 80 %, and 70 %). When a prediction region has 0 uncertain labels, the size of the prediction region is 1 (contains a single multi-label prediction). When we have 1 uncertain label, the prediction region size is 2, since the region contains a multi-label prediction for each of the 2 possible values of the uncertain binary label. Generally, for n uncertain binary labels, the prediction region size is 2^n. The error rates presented in Fig. 1 demonstrate the validity of the BR-MLCP, since they are always below the rate given by the confidence level. Thus, we demonstrate the ability to control the error rate of BR-MLCP. Nevertheless, when we have a high confidence level, we lose some certainty in the predictions. In the figure, we can see that for 95 % level of confidence the number of certain predictions is 0, and a significant percentage of predictions contained all 6 labels as uncertain labels. The algorithm provides uncertain results when there is not enough information to give a single result for a given confidence level.

It is admitted that for a multi-label problem, the error measure that was defined in (8) is strict. Nonetheless, if we lower the confidence level, we get more certainty in the predictions. For example, at 80 % and 70 % levels of confidence, we have a significant amount of prediction regions with less uncertain labels.

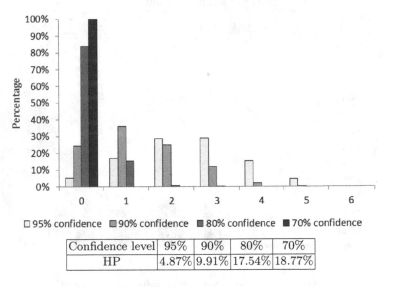

Confidence level	95%	90%	80%	70%
HP	4.87%	9.91%	17.54%	18.77%

Fig. 2. Percentages of prediction regions with number of uncertain labels for different levels of confidence, and their respective hamming loss on the Emotions dataset.

In Fig. 2, we provide the results of the BR-MLCP using (18). Here the error measure is less strict, and thus we get satisfactory certainty in our prediction regions. The error is measured in terms of hamming loss. As shown in the figure, the hamming loss in the prediction regions does not exceed the allowed rate given by the confidence level. Therefore, we demonstrate that the BR-MLCP can control the hamming loss in the prediction regions and provide useful prediction regions. Additionally, the hamming loss at 70 % confidence does not exceed 18.77 %. We also notice that at this confidence level, we have 100 % certain predictions.

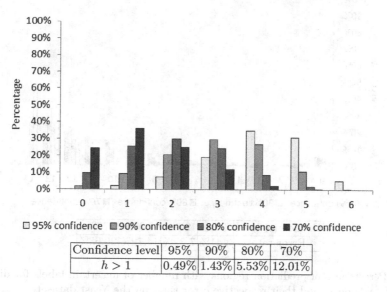

Confidence level	95%	90%	80%	70%
$h > 1$	0.49%	1.43%	5.53%	12.01%

Fig. 3. Percentages of prediction regions with number of uncertain labels for different levels of confidence, and their respective error rates (using (13) with $h = 1$) on the Emotions dataset.

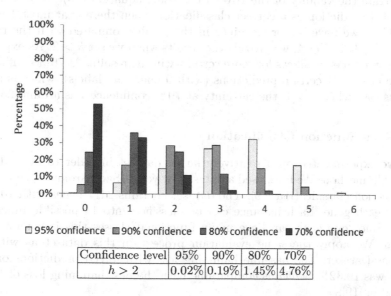

Confidence level	95%	90%	80%	70%
$h > 2$	0.02%	0.19%	1.45%	4.76%

Fig. 4. Percentages of prediction regions with number of uncertain labels for different levels of confidence, and their respective error rates (using (13) with $h = 2$) on the Emotions dataset.

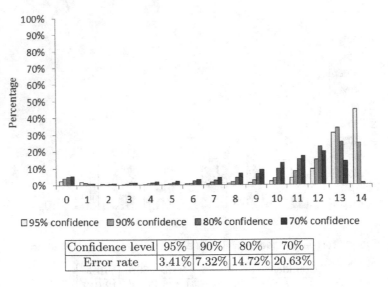

Confidence level	95%	90%	80%	70%
Error rate	3.41%	7.32%	14.72%	20.63%

Fig. 5. Percentages of prediction regions with number of uncertain labels for different levels of confidence, and their respective error rates on the Yeast dataset.

In Fig. 3, we provide the results of the BR-MLCP using (13). Here, we have an error when the hamming loss h of a prediction region exceeds 1. The results demonstrate the validity of the BR-MLCP using equation (13). We consider a multi-label prediction as a correct classification when there is at most 1 wrong label. Thus, we have better certainty in the results compared with the results given in Fig. 1. In Fig. 4, we provide the results when we set $h > 2$. As expected, this less strict metric allows for more certainty in the results. At 70 % confidence, we have near 50 % certain predictions (with 0 uncertain labels), whereas in the previous case when $h > 1$, the certainty at 70 % confidence was around 25 %.

4.2   Gene Function Classification

We have experimented on a relatively larger dataset in order to evaluate the BR-MLCP method. We have used a dataset for yeast (Saccharomyces cerevisiae) gene function classification [19]. The dataset contains 2417 genes with 103 features in each gene. Each instance can be classified into 14 possible functional groups. Since one gene can have many functional classes this is a multi-label problem. We apply the same evaluation process on this dataset as with the Emotions Dataset. The baseline hamming loss with forced predictions on this dataset was 19.32 %, which is comparable with the best hamming loss of 19.5 % reported in [19].

In Fig. 5, we provide the results of the BR-MLCP using (8). As shown in the figure, the percentage of prediction regions which contained a certain multi-label prediction is near 5 %. This is true for any given confidence level. As explained

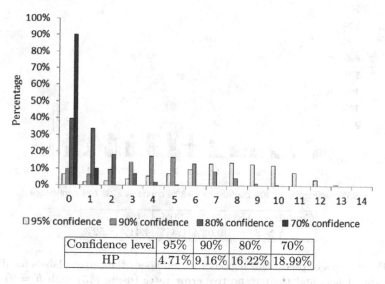

Confidence level	95%	90%	80%	70%
HP	4.71%	9.16%	16.22%	18.99%

Fig. 6. Percentages of prediction regions with number of uncertain labels for different levels of confidence, and their respective hamming loss on the Yeast dataset.

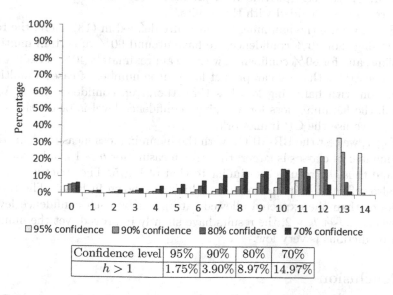

Confidence level	95%	90%	80%	70%
$h > 1$	1.75%	3.90%	8.97%	14.97%

Fig. 7. Percentages of prediction regions with number of uncertain labels for different levels of confidence, and their respective error rates (using (13) with $h = 1$) on the Yeast dataset.

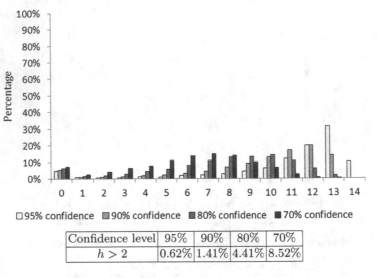

Confidence level	95%	90%	80%	70%
$h > 2$	0.62%	1.41%	4.41%	8.52%

Fig. 8. Percentages of prediction regions with number of uncertain labels for different levels of confidence, and their respective error rates (using (13) with $h = 2$) on the Yeast dataset.

previously, using (8) as an error measure can be very strict for multi-label problems. This becomes more clear when the number of classes is larger. Nevertheless, the BR-MLCP can still provide valid prediction regions, as it is demonstrated by the error rates provided with the results.

In Fig. 6, we use the hamming loss measure defined in (18). Here, the results are promising. For 70 % confidence, we have around 90 % of certain multi-label predictions, and for 80 % confidence, we have approximately 40 % certainty rates. This demonstrates that we can predict for a given number of cases a multi-label classification with hamming loss less than the given confidence level. As it is expected, the hamming loss for any given confidence level is below the allowed rate, since we use the CP framework.

In Fig. 7, we test the BR-MLCP with the hamming loss measure defined (13). As the number of classes is larger, the error measure for $h > 1$ can be considered strict, and thus the results are similar to that of Fig. 5. The strictness of $h > 1$ loss is also reflected on the error rates which are shown in Fig. 7. The rates are much lower than the expected allowed rate given by each confidence level. In Fig. 8 where we set $h > 2$, the results have slightly improved, yet the number of certain predictions is very low.

5   Conclusion

The CP framework provides reliable measures of confidence to predictions of Machine Learning algorithms. We gave an overview of CP and the extended MLCP framework. We have applied the BR-MLCP algorithm on two multi-label datasets: one for classifying music into emotions, and another for Yeast gene function classification. We have experimented with three measures of error. Hamming loss, which is a widely used measure of error for multi-label problems, was shown to give more informative prediction regions in the sense that the size of the prediction regions was much smaller. Our defined hamming loss confidence measure has allowed us to control the hamming loss in our predictions, while keeping the error rates below or near the pre-set level. In the future, we would like to conduct more experiments with a variety of hamming loss values for the proposed confidence measure. Additionally, we would like to examine the use of other techniques as underlying algorithms combined with ways of addressing the class imbalance problem resulting from the BR transformation that was used in our work.

Acknowledgements. The authors would like to thank Dr. Ilia Nouretdinov and Professor Volodya Vovk for their helpful discussions and indications. We would also like to thank the anonymous reviewers for their constructive comments.

References

1. Vovk, V., Gammerman, A., Shafer, G.: Algorithmic Learning in a Random World. Springer, New York (2005)
2. Gammerman, A., Vovk, V., Vapnik, V.: Learning by transduction. In: Uncertainty in Artificial Intelligence, pp. 148–155. Morgan Kaufmann (1998)
3. Balasubramanian, V.N., Ho, S.-S., Vovk, V. (eds.): Conformal Prediction for Reliable Machine Learning. Morgan Kaufmann, Boston (2014). doi:10.1016/B978-0-12-398537-8.00014-6. http://www.sciencedirect.com/science/article/pii/B9780123985378000146. ISBN 978-0-12-398537-8
4. Saunders, C., Gammerman, A., Vovk, V.: Transduction with confidence and credibility. In: Proceedings of the 16th International Joint Conference on Artificial Intelligence, vol. 2, pp. 722–726. Morgan Kaufmann, Los Altos (1999)
5. Papadopoulos, H., Vovk, V., Gammerman, A.: Regression conformal prediction with nearest neighbours. J. Artif. Intell. Res. **40**, 815–840 (2011)
6. Devetyarov, D., Nouretdinov, I.: Prediction with confidence based on a random forest classifier. In: Papadopoulos, H., Andreou, A.S., Bramer, M. (eds.) AIAI 2010. IFIP AICT, vol. 339, pp. 37–44. Springer, Heidelberg (2010). doi:10.1007/978-3-642-16239-8_8
7. Bellotti, T., Luo, Z., Gammerman, A., Van Delft, F.W., Saha, V.: Qualified predictions for microarray and proteomics pattern diagnostics with confidence machines. Int. J. Neural Syst. **15**(4), 247–258 (2005)
8. Gammerman, A., Nouretdinov, I., Burford, B., Chervonenkis, A., Vovk, V., Luo, Z.: Clinical mass spectrometry proteomic diagnosis by conformal predictors. Stat. Appl. Genet. Mol. Biol. **7**(2) (2008). http://dx.doi.org/10.2202/1544-6115.1385
9. Papadopoulos, H., Gammerman, A., Vovk, V.: Confidence predictions for the diagnosis of acute abdominal pain. In: Iliadis, Maglogiann, Tsoumakasis, Vlahavas, Bramer, (eds.) Artificial Intelligence Applications and Innovations III. IFIP, vol. 296, pp. 175–184. Springer, US (2009)
10. Papadopoulos, H., Gammerman, A., Vovk, V.: Reliable diagnosis of acute abdominal pain with conformal prediction. Int. J. Eng. Intell. Syst. Electr. Eng. Commun. **17**(2–3), 127–137 (2009). ISSN 1472–8915

11. Balasubramanian, V.N., Chakraborty, S., Panchanathan, S.: Conformal predictions for information fusion. Annals Math. Artif. Intell. **74**(1–2), 45–65 (2015). doi:10. 1007/s10472-013-9392-4. ISSN 1012–2443
12. Yang, M., Nouretdunov, I., Luo, Z., Gammerman, A.: Feature selection by conformal predictor. In: Iliadis, L., Maglogiannis, I., Papadopoulos, H. (eds.) EANN/AIAI 2011, Part II. IFIP AICT, vol. 364, pp. 439–448. Springer, Heidelberg (2011). doi:10.1007/978-3-642-23960-1_51. ISBN 978-3-642-23959-5
13. Tsoumakas, G., Katakis, I.: Multi-label classification: an overview. Int. J. Data Warehouse. Min. **1–13**, 2007 (2007)
14. Wieczorkowska, A., Synak, P., Raś, Z.W.: Multi-label classification of emotions in music. In: Kłopotek, M.A., Wierzchoń, S.T., Trojanowski, K. (eds.) Intelligent Information Processing and Web Mining. Advances in Soft Computing, vol. 35, pp. 307–315. Springer, Heidelberg (2006). doi:10.1007/3-540-33521-8_30. ISBN 978-3-540-33520-7
15. Papadopoulos, H.: A cross-conformal predictor for multi-label classification. In: Iliadis, L., Maglogiannis, I., Papadopoulos, H., Sioutas, S., Makris, C. (eds.) Artificial Intelligence Applications and Innovations. IFIP AICT, vol. 437, pp. 241–250. Springer, Heidelberg (2014). doi:10.1007/978-3-662-44722-2_26. ISBN 978-3-662-44721-5
16. Wang, H., Liu, X., Lv, B., Yang, F., Hong, Y.: Reliable multi-label learning via conformal predictor, random forest for syndrome differentiation of chronic fatigue in traditional chinese medicine. PLoS ONE **9**(6), e99565 (2014)
17. Wang, H., Liu, X., Nouretdinov, I., Luo, Z.: A comparison of three implementations of multi-label conformal prediction. In: Gammerman, A., Vovk, V., Papadopoulos, H. (eds.) Statistical Learning and Data Sciences. LNCS, vol. 9047, pp. 241–250. Springer, Switzerland (2015). doi:10.1007/978-3-319-17091-6_19. ISBN 978-3-319-17091-6
18. Papadopoulos, H., Vovk, V., Gammerman, A.: Qualified predictions for large data sets in the case of pattern recognition. In: Proceedings of the International Conference on Machine Learning and Applications (ICMLA 2002), pp. 159–163. CSREA Press (2002)
19. Elisseeff, A., Weston, J.: A kernel method for multi-labelled classification. In: Advances in Neural Information Processing Systems 14, pp. 681–687. MIT Press (2001)

A Metric to Improve the Robustness of Conformal Predictors in the Presence of Error Bars

Andrea Murari[1], Saeed Talebzadeh[2], Jesús Vega[3],
Emmanuele Peluso[2(✉)], Michela Gelfusa[2], Michele Lungaroni[2],
and Pasqualino Gaudio[2]

[1] Consorzio RFX (CNR, ENEA, INFN), Acciaierie Venete SpA,
Universita' di Padova, Corso Stati Uniti 4, 35127 Padua, Italy
murari@igi.cnr.it
[2] University of Rome "Tor Vergata", Via del Politecnico 1, 00133 Rome, Italy
talebzadeh.saeed@gmail.com,
{emmanuele.peluso,michele.lungaroni}@uniroma2.it,
{gelfusa,gaudio}@ing.uniroma2.it
[3] Asociación EURATOM/CIEMAT para Fusión, 28040 Madrid, Spain
jesus.vega@ciemat.es

Abstract. Conformal predictors, currently applied to many problems in various fields determine precise levels of confidence in new predictions on the basis only of the information present in the past data, without making recourse to any assumptions except that the examples are generated independently from the same probability distribution. In this paper, the robustness of their results is assessed for the cases in which the data are affected by error bars. This is the situation typical of the physical sciences, whose data are often the results of complex measurement procedures, unavoidably affected by noise. Assuming the noise presents a normal distribution, the Geodesic Distance on Gaussian Manifolds provides a statistical principled and quite effective method to handle the uncertainty in the data. A series of numerical tests prove that adopting this metric in conformal predictors improves significantly their performance, compared to the Euclidean distance, even for relatively low levels of noise.

Keywords: Conformal predictors · Geodesic distance · Inference methods · Error bars

1 Conformal Predictors and Measurement Errors

Machine-learning methods work often very well and have found many applications in both the public and the private sector. On the other hand, the reliability of their performance is typically proven asymptotically and is therefore not very useful in practice. Conformal predictors, which perform competitively in terms of success rates, include from their conception simple and useful measures of confidence [1]. Conformal prediction can be based on any technique of point prediction for classification or regression, including support-vector machines, decision trees, neural networks and Bayesian

© Springer International Publishing Switzerland 2016
A. Gammerman et al. (Eds.): COPA 2016, LNAI 9653, pp. 105–115, 2016.
DOI: 10.1007/978-3-319-33395-3_8

methods. Starting from the point prediction tool, the conformal predictor consists of building a nonconformity measure, which determines how unusual an example is relative to previous examples. The conformal algorithm, based on the statistical concept of the p-values, turns this nonconformity measure into prediction regions. Given a nonconformity measure, the conformal algorithm produces a prediction region Uε for every probability of error ε. The region Uε is a $(1-ε)$-prediction region; it classifies the next example with probability at least $1-ε$. Therefore conformal predictors are conservatively valid, which means that the probability they make a mistake when their output is at confidence level $1-ε$ is not greater than ε.

In most of the non-conformity measures utilised by conformal predictors, the Euclidean distance is implicitly assumed to be the proper metric to adopt in the calculation of the non-conformity measure and the p-values. The Euclidean distance has a precise geometrical meaning and a very long historical pedigree. However, it implicitly requires considering all data as single infinitely precise values. This assumption can be appropriate in other applications but it is obviously not the case in physics, since all the measurements typically present an error bar. An alternative idea is to use a new distance between data, which would take into account the measurement uncertainties. The causes of uncertainties in the measurements are typically many, which from a statistical point of view can be considered random variables. As a consequence, their global contribution can be often modelled as a noise of normal distribution. The idea, behind the approach proposed in this paper, consists therefore of considering the measurements not as points, but as Gaussian distributions [2]. Modelling measurements not as point values, but as Gaussian distributions, requires defining a distance between Gaussians. This distance must be the Geodesic on the Gaussian Manifold (GDGM) of the measurements and can be expressed as a closed formula (see Sect. 3) [3]. As shown in the rest of the paper, adopting this geodesic distance can increase significantly the accuracy of traditional conformal predictors, even when the data are affected by a very limited level of noise.

With regard to the structure of the paper, next Section provides a short introduction to the general framework of conformal prediction. The mathematical background to the main mathematical tool introduced in the paper: the Geodesic Distance on Gaussian Manifolds, is the subject of Sect. 3. The proposed method is assessed with a series of numerical tests using a toy model described in Sect. 4. Section 5 reports in detail the results of the numerical tests. Conclusions and lines of future work are provided in the last Section of the paper.

2 The Framework of Conformal Prediction for Classification

The task of classification basically consists of attributing objects to different classes. Mathematically this can be formalised by considering successive ordered pairs (x_1, y_1), (x_2, y_2).......which are called examples. Each example consists of an object x_i and its label y_i, where the former represents the feature vector that describes the object i. The objects are elements of a measurable space X called the *object space*; the labels are elements of a measurable space Y called the *label space*. It is common practice to adopt a more compact notation, according to which z_i indicate the ordered pair (x_i, y_i), and $Z := X \times Y$ is defined as the *example space*.

Many machine learning tools are available to perform classification. On the other hand, as mentioned earlier, the vast majority of them cannot easily quantify the quality of their predictions. On the contrary, conformal predictors have been conceived explicitly to quantify the reliability of their predictions. They achieve this on the basis of the past examples. To this end, for each new sample to classify, it is necessary to measure how different the new one is from the old examples. In this perspective, a nonconformity measure is defined, which allows calculating a nonconformity score to estimate how different a new example is from a bag of old ones. A bag of size $n \in N$ is a collection of n elements some of which may be identical. In this paper, the notation $< z_1,..., z_n >$ indicates a bag of n elements.

Given a nonconformity measure A and a bag $< z_1,...,z_n >$, the nonconformity score can be calculated as:

$$\alpha_i := A(\langle z_1, \ldots, z_{i-1}, z_{i+1}, \ldots, z_n \rangle, z_i) \qquad (1)$$

for each example z_i in the bag. Because nonconformity measures are not absolute but relative, the numerical value of α_i does not, by itself, determines how unusual z_i is according to the measure A. To really quantify how unusual a sample is, it is necessary to compare α_I with the nonconformity measures α_j of the other members of the bag. The p-value is a convenient and statistically sound way of calculating how anomalous a new example is. By definition the p-value is the fraction

$$Pval = \frac{\#\{j = 1, \ldots, n : \alpha_j \geq \alpha_i\}}{n} \qquad (2)$$

This indicator, which lies between $1/n$ and 1, is the fraction of the examples in the bag as non conforming as z_i and in literature is called p-value of the element z_i ($p_{val}(z_i)$). The symbol "#" stands in fact for the number of elements "j" in the collection having a nonconformity score higher or at least the same nonconformity of the element "i". The lower the p-value, i.e. the closer to its lower bound $1/n$ ("j" includes "i" in fact) for

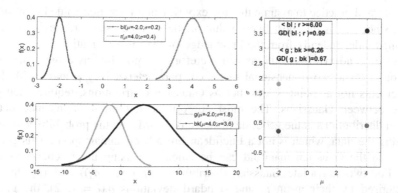

Fig. 1. Examples to illustrate how the GDGM determines the distance between two Gaussians. The two couples of pdf in the figure have the same mean but different σ. The geodesic distance between the two with higher σ is much smaller.

large n, the more non conforming z_i is and the more likely it can be considered as an outlier; this means that z_i is not representative of the typical member of the bag. If the p-value is large, i.e. close to its upper bound 1, then z_i is very conforming or very representative of the typical member of the bag. The new sample is attributed to the class with the highest p-value.

On the basis of the p-values, conformal predictors allow calculating, for each new classification, two indicators, confidence and credibility, which quantify the reliability of the prediction. Credibility is defined as the largest p-value; confidence is defined as 1-2nd largest p-value. Confidence can be interpreted as the probability that the prediction, corresponding to the maximal p-value, is correct. A low credibility, typically less than 0.05, intuitively means that either the training set is non random or the test object is not representative of the training set. If the maximum p-value appears in more than one class, an ambiguity is present and the algorithm is not able to classify the sample. It is important to emphasize that confidence and credibility of the prediction play an analogous role to the observed level of significance in statistical parameter tests.

3 Geodesic Distance on Gaussian Manifolds

As mentioned in the previous section, in the natural sciences the data available are typically the result of experimental measurements. In this context, all measurements are affected by uncertainties referred to as error bars. The sources of this uncertainty are normally quite many and therefore it is more than reasonable to assume that the pdf of the noise is normal. Each measurement can therefore be modelled as a probability density function (pdf) of the Gaussian type, determined by its mean μ and its standard deviation σ:

$$p(x; \mu, \sigma) = \frac{1}{\sigma\sqrt{2\pi}} exp\left[-\frac{(x-\mu)^2}{2\sigma^2}\right] \tag{3}$$

It is normal practice to assume that the experimental measured value is the mean of the pdf, since this is the most likely value of the pdf. The standard deviation can be determined independently from the knowledge of the instrumentation.

The set of normal distributions can therefore be modelled as a two dimensional space, or better a two dimensional manifold, parameterized by μ and σ. Modelling measurements not as point values, but as Gaussian distributions, requires defining a distance between Gaussians. The most appropriate definition of distance between Gaussian distributions is the geodesic distance (GDGM), on the probabilistic manifold containing the data, which is not a Euclidean but a Riemannian space. This geodesic distance on the Gaussian manifold can be calculated using the Fischer-Rao metric [3, 4]. For two univariate Gaussian distributions $(p_1(x|\mu_1, \sigma_1))$ and $(p_2(x|\mu_2, \sigma_2))$, parameterised by their mean μ_i and standard deviations $\sigma_i (i = 1, 2)$, the geodesic distance GDGM is given by:

$$GD(p_1 \| p_2) = \sqrt{2} ln \frac{1+\delta}{1-\delta} = 2\sqrt{2} tanh^{-1}\delta, where\ \delta = [\frac{(\mu_1 - \mu_2)^2 + 2(\sigma_1 - \sigma_2)^2}{(\mu_1 - \mu_2)^2 + 2(\sigma_1 + \sigma_2)^2}]^{\frac{1}{2}}$$

$$(4)$$

As will be shown in detail in the next sections, the replacement of the Euclidean distance with the GDGM improves significantly the robustness of the classification compared to the case of the Euclidean distance. In Fig. 1 a graphical example of the improvement obtained using the metric in Eq. (4) is shown. Considering a Cartesian coordinate system (μ, σ), where each point represents a Gaussian distribution, the Euclidean distance between the four points, so between the four distributions, is higher between the two wider distributions. On the other hand, considering the Geodesic Distance, the lower distance is obtained considering the wider distributions. This behaviour reflects the physical interpretation according to which physical quantities having higher error bars, are to be considered closer and more similar than those with narrower error bars.

4 A Toy Model

To exemplify and prove the usefulness of the method proposed in this paper, a series of numerical test has been performed. They are based on a toy model already introduced in [5]. The simplicity of the model allows appreciating both the nature of the problem and the advantages of adopting the proposed metric, the GDGM. The classification task consists of classifying points on a straight line, on which three classes have been defined. The problem is represented graphically in Fig. 2. The aim is to classify the new point Q with confidence and credibility.

For the purpose of this example, the classification is based on the nearest neighbour. Mathematically, given a "bag" $\{z_1,...,z_{n-1}\}$, where each z_i consists of a feature vector x_i and a non-numerical label y_i, when a new example $z_n = (x_n, y_n)$ becomes available for classification, its feature vector x_n is known but its label y_n is not. The nearest-neighbour method finds the x_i closest to x_n and its label y_i becomes the prediction of y_n. A natural way to measure the nonconformity of the new example z_n with respect to the old examples z_i consists of comparing x's distances to old objects with the same label to its distance to old objects with a different label. For example, the nonconformity scores can be defined as:

$$\alpha_i = \frac{d_{sl}}{d_{dl}}$$

$$(5)$$

$$d_{sl} = \min\{|x_j - x_i| : 1 \leq j \leq n\ \&\ j \neq i\ \&\ y_j = y_i\}$$

$$d_{dl} = \min\{|x_j - x_i| : 1 \leq j \leq n\ \&\ j \neq i\ \&\ y_j \neq y_i\}$$

For the new point $Q = 14.85$ shown in Fig. 2 (the non-conformity measurement is presented in Table 1), therefore the P values credibility and confidence assume the values 1 and 0.9844, respectively; and point Q belongs to Class C.

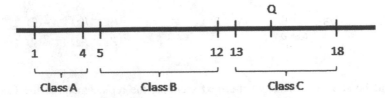

Fig. 2. The toy model. A new point Q must be classified as belonging to class A, B or C.

Table 1. Non-conformity measurements for point Q = 14.85

Object i	α if Y = A	α if Y = B	α if Y = C	Object i	α if Y = A	α if Y = B	α if Y = C
1	0.0625	0.0625	0.0625	10	0.0833	0.0833	0.0833
1.25	0.0667	0.0667	0.0667	10.25	0.0909	0.0909	0.0909
1.5	0.0714	0.0714	0.0714	10.5	0.1	0.1	0.1
1.75	0.0769	0.0769	0.0769	10.75	0.1111	0.1111	0.1111
2	0.0833	0.0833	0.0833	11	0.125	0.125	0.125
2.25	0.0909	0.0909	0.0909	11.25	0.1429	0.1429	0.1429
2.5	0.1	0.1	0.1	11.5	0.1667	0.1667	0.1667
2.75	0.1111	0.1111	0.1111	11.75	0.2	0.2	0.2
3	0.125	0.125	0.125	12	0.25	0.25	0.25
3.25	0.1429	0.1429	0.1429	13	0.25	0.25	0.25
3.5	0.1667	0.1667	0.1667	13.25	0.2	0.2	0.2
3.75	0.2	0.2	0.2	13.5	0.1852	0.1852	0.1667
4	0.25	0.25	0.25	13.75	0.2273	0.2273	0.1429
5	0.25	0.25	0.25	14	0.2941	0.2941	0.125
5.25	0.2	0.2	0.2	14.25	0.4167	0.4167	0.1111
5.5	0.1667	0.1667	0.1667	14.5	0.7143	0.7143	0.1
5.75	0.1429	0.1429	0.1429	14.75	2.5	2.5	0.0364
6	0.125	0.125	0.125	15	1.6667	1.6667	0.05
6.25	0.1111	0.1111	0.1111	15.25	0.625	0.625	0.0769
6.5	0.1	0.1	0.1	15.5	0.3846	0.3846	0.0714
6.75	0.0909	0.0909	0.0909	15.75	0.2778	0.2778	0.0667
7	0.0833	0.0833	0.0833	16	0.2174	0.2174	0.0625
7.25	0.0769	0.0769	0.0769	16.25	0.1786	0.1786	0.0588
7.5	0.0714	0.0714	0.0714	16.5	0.1515	0.1515	0.0556
7.75	0.0667	0.0667	0.0667	16.75	0.1316	0.1316	0.0526
8	0.0625	0.0625	0.0625	17	0.1163	0.1163	0.05
8.25	0.0588	0.0588	0.0588	17.25	0.1042	0.1042	0.0476
8.5	0.0556	0.0556	0.0556	17.5	0.0943	0.0943	0.0455
8.75	0.0588	0.0588	0.0588	17.75	0.0862	0.0862	0.0435
9	0.0625	0.0625	0.0625	18	0.0794	0.0794	0.0417
9.25	0.0667	0.0667	0.0667	14.85	108.5	...	...
9.5	0.0714	0.0714	0.0714	14.85	...	28.5	...
9.75	0.0769	0.0769	0.0769	14.85	...	...	0.0351
Continue -------->				p-value	0.0156	0.0156	1

In the previous example, the conformity measure of Eq. (5) has been calculated using the Euclidean distance between the various points. All the derived quantities are therefore also based on this metric. In the case of measurements affected by noise, the Euclidean metric is not adequate and adopting the GDGM provides several improvements as discussed in the next section.

5 Results of the Numerical Tests

In order to assess the potential of the GDGM metric to counteract the effect of noise, a series of systematic tests has been performed using the toy model introduced in the previous section. To this end, a series of points have been automatically generated along the straight line of Fig. 2. These are to be considered the right values of the physical quantity to measure. Then Gaussian noise, with zero mean and standard deviations equals to a percentage (10 %,20 %,...) of the value itself, has been added to the previously generated points. Adding this noise to the data provides the actual values to be considered as the available measurements, affected by additive noise of Gaussian distribution. These points have been then classified with the nonconformity measure based on the next neighbour criterion using both the Euclidean distance and the GDGM. The results have been reported in Table 2 for the Euclidean distance as metric and in Table 3 for the GDGM as metric.

Table 2. Classification using the Euclidean distance to calculate the nearest neighbour. The first column reports the accuracy (Acc.); the second the credibility (Cred.) and the third the confidence (Conf.). The following column reports the same quantities but for different levels of noise. The top of the table reports the average values for all the 50 points.

	Euclidean							
Noise=10%			Noise=20%			Noise=30%		
Acc.	Cred.	Conf.	Acc.	Cred.	Conf.	Acc.	Cred.	Conf.
Average values for all runs:								
84.67	0.80	0.89	72.77	0.77	0.86	65.03	0.76	0.84
Average values for each run:								
87.18	0.84	0.98	71.79	0.71	0.83	69.23	0.81	0.81
76.92	0.77	0.81	79.49	0.82	0.91	69.23	0.76	0.83
89.74	0.81	0.93	79.49	0.78	0.86	58.97	0.71	0.76
89.74	0.78	0.88	66.67	0.83	0.86	53.85	0.72	0.73
79.49	0.80	0.93	58.97	0.74	0.78	71.79	0.69	0.83
84.62	0.83	0.93	74.36	0.74	0.86	64.10	0.75	0.81
84.62	0.77	0.91	82.05	0.78	0.91	71.79	0.76	0.81
84.62	0.81	0.96	66.67	0.64	0.81	53.85	0.70	0.76
92.31	0.81	0.93	66.67	0.82	0.88	58.97	0.81	0.83
79.49	0.78	0.81	82.05	0.79	0.91	66.67	0.73	0.83
79.49	0.81	0.88	76.92	0.80	0.86	69.23	0.80	0.86

Table 2. (*Continued*)

87.18	0.84	0.96	87.18	0.83	0.96	66.67	0.75	0.86
87.18	0.78	0.88	61.54	0.68	0.76	66.67	0.75	0.88
87.18	0.79	0.91	69.23	0.87	0.91	64.10	0.86	0.91
82.05	0.78	0.88	61.54	0.81	0.83	69.23	0.86	0.91
82.05	0.76	0.86	69.23	0.72	0.78	61.54	0.74	0.81
74.36	0.76	0.86	64.10	0.81	0.88	69.23	0.89	0.93
74.36	0.80	0.83	79.49	0.80	0.88	64.10	0.74	0.86
82.05	0.80	0.86	82.05	0.79	0.86	56.41	0.84	0.88
87.18	0.78	0.88	76.92	0.83	0.91	74.36	0.72	0.81
87.18	0.75	0.86	66.67	0.77	0.88	64.10	0.74	0.81
89.74	0.88	0.96	69.23	0.81	0.88	71.79	0.83	0.86
79.49	0.83	0.86	74.36	0.71	0.86	61.54	0.64	0.83
92.31	0.81	0.88	66.67	0.75	0.83	66.67	0.78	0.88
87.18	0.87	0.96	82.05	0.75	0.88	69.23	0.71	0.83
84.62	0.78	0.88	69.23	0.83	0.86	66.67	0.75	0.83
79.49	0.82	0.91	74.36	0.74	0.91	58.97	0.72	0.78
79.49	0.80	0.83	69.23	0.73	0.83	69.23	0.83	0.91
84.62	0.75	0.93	74.36	0.80	0.88	76.92	0.82	0.91
87.18	0.84	0.88	66.67	0.80	0.86	58.97	0.65	0.78
92.31	0.81	0.96	76.92	0.82	0.91	66.67	0.71	0.78
94.87	0.83	0.96	84.62	0.86	0.88	71.79	0.75	0.86
76.92	0.75	0.78	69.23	0.76	0.88	64.10	0.76	0.83
84.62	0.86	0.91	76.92	0.82	0.91	51.28	0.65	0.76
79.49	0.79	0.86	79.49	0.76	0.91	58.97	0.75	0.78
87.18	0.81	0.93	74.36	0.75	0.88	69.23	0.80	0.88
87.18	0.85	0.91	74.36	0.78	0.81	61.54	0.68	0.73
84.62	0.75	0.91	69.23	0.72	0.76	58.97	0.71	0.76
79.49	0.79	0.86	79.49	0.76	0.91	58.97	0.75	0.78
87.18	0.81	0.93	74.36	0.75	0.88	69.23	0.80	0.88
87.18	0.85	0.91	74.36	0.78	0.81	61.54	0.68	0.73
84.62	0.75	0.91	69.23	0.72	0.76	58.97	0.71	0.76
89.74	0.77	0.88	66.67	0.74	0.86	79.49	0.81	0.88
87.18	0.78	0.86	76.92	0.70	0.86	76.92	0.78	0.86
92.31	0.85	0.96	64.10	0.80	0.86	69.23	0.80	0.88
69.23	0.65	0.76	71.79	0.75	0.86	69.23	0.76	0.86
84.62	0.83	0.88	74.36	0.78	0.81	61.54	0.81	0.91
79.49	0.65	0.78	79.49	0.78	0.88	69.23	0.76	0.91
89.74	0.80	0.96	71.79	0.80	0.88	64.10	0.78	0.91
79.49	0.79	0.81	61.54	0.67	0.73	51.28	0.75	0.83
89.74	0.87	0.93	76.92	0.84	0.88	64.10	0.76	0.86
87.18	0.80	0.86	64.10	0.66	0.76	51.28	0.75	0.81
84.62	0.86	0.91	79.49	0.73	0.88	76.92	0.79	0.86
89.74	0.78	0.91	76.92	0.81	0.88	51.28	0.68	0.81

Table 3. Classification using the Geodesic Distance on Gaussian Manifolds distance to calculate the nearest neighbour. The first column reports the accuracy (Acc.); the second the credibility (Cred.) and the third the confidence (Conf.). The following column reports the same quantities but for different levels of noise. The top of the table reports the average values for all the 50 points.

| | Geodesic | | | | | | | |
| Noise=10% | | | Noise=20% | | | Noise=30% | | |
Acc.	Cred.	Conf.	Acc.	Cred.	Conf.	Acc.	Cred.	Conf.
Average values for all runs:								
89.33	**0.75**	**0.95**	**80.67**	**0.71**	**0.94**	**74.46**	**0.68**	**0.93**
Average values for each run:								
87.18	0.79	0.98	79.49	0.71	0.91	82.05	0.76	0.93
89.74	0.74	0.93	87.18	0.75	0.98	79.49	0.63	0.96
94.87	0.75	0.98	87.18	0.70	0.93	71.79	0.74	0.91
92.31	0.76	0.91	74.36	0.65	0.93	64.10	0.57	0.83
82.05	0.76	0.98	76.92	0.72	0.98	74.36	0.62	0.93
87.18	0.78	0.96	82.05	0.71	0.93	79.49	0.65	0.96
89.74	0.72	0.96	87.18	0.75	0.96	87.18	0.73	0.96
87.18	0.77	0.98	76.92	0.65	0.91	69.23	0.63	0.91
92.31	0.75	0.96	71.79	0.71	0.93	71.79	0.74	0.96
89.74	0.79	0.91	87.18	0.70	0.96	82.05	0.70	0.98
89.74	0.79	0.98	89.74	0.81	0.98	76.92	0.69	0.93
89.74	0.73	0.98	89.74	0.70	0.98	76.92	0.69	0.96
94.87	0.74	0.96	76.92	0.68	0.91	66.67	0.63	0.86
89.74	0.77	0.93	71.79	0.66	0.93	69.23	0.65	0.96
87.18	0.75	0.93	69.23	0.64	0.91	71.79	0.67	0.93
92.31	0.73	0.96	82.05	0.73	0.91	71.79	0.71	0.91
82.05	0.72	0.96	69.23	0.66	0.93	71.79	0.70	0.96
84.62	0.77	0.93	84.62	0.72	0.98	71.79	0.64	0.93
89.74	0.76	0.93	89.74	0.79	0.93	66.67	0.79	0.98
92.31	0.75	0.93	84.62	0.72	0.98	89.74	0.78	0.96
92.31	0.78	0.91	71.79	0.72	0.93	74.36	0.69	0.91
92.31	0.79	0.98	76.92	0.69	0.96	82.05	0.79	0.96
87.18	0.73	0.93	82.05	0.72	0.93	66.67	0.62	0.91
97.44	0.75	0.93	82.05	0.75	0.98	74.36	0.62	0.96
87.18	0.76	0.96	89.74	0.70	0.96	82.05	0.71	0.96

Table 3. *(Continued)*

92.31	0.75	0.96	74.36	0.66	0.91	79.49	0.72	0.96
82.05	0.74	0.93	74.36	0.64	0.93	76.92	0.69	0.96
82.05	0.76	0.91	76.92	0.66	0.91	69.23	0.60	0.91
89.74	0.68	0.98	74.36	0.66	0.93	84.62	0.67	0.98
92.31	0.78	0.93	74.36	0.70	0.93	71.79	0.63	0.88
89.74	0.77	0.98	82.05	0.67	0.96	76.92	0.69	0.88
94.87	0.73	0.96	89.74	0.69	0.93	79.49	0.66	0.93
87.18	0.73	0.93	76.92	0.65	0.96	69.23	0.66	0.88
87.18	0.75	0.93	82.05	0.67	0.96	61.54	0.56	0.88
84.62	0.72	0.91	84.62	0.70	0.96	71.79	0.69	0.91
92.31	0.74	0.98	79.49	0.74	0.93	74.36	0.64	0.91
87.18	0.72	0.93	87.18	0.75	0.93	74.36	0.70	0.86
87.18	0.73	0.93	84.62	0.67	0.93	71.79	0.63	0.88
92.31	0.74	0.96	71.79	0.70	0.93	87.18	0.72	0.96
89.74	0.74	0.93	84.62	0.69	0.93	82.05	0.68	0.91
94.87	0.73	0.98	74.36	0.76	0.96	74.36	0.63	0.93
82.05	0.71	0.91	79.49	0.71	0.93	79.49	0.71	0.93
87.18	0.76	0.91	82.05	0.74	0.96	66.67	0.73	0.96
87.18	0.68	0.88	89.74	0.80	0.98	71.79	0.66	0.91
92.31	0.74	0.98	79.49	0.75	0.96	69.23	0.61	0.96
89.74	0.76	0.93	71.79	0.61	0.83	64.10	0.72	0.93
94.87	0.77	0.98	87.18	0.75	0.98	76.92	0.72	0.98
87.18	0.74	0.96	76.92	0.65	0.91	66.67	0.65	0.96
87.18	0.78	0.93	87.18	0.72	0.96	84.62	0.75	0.93
92.31	0.76	0.98	87.18	0.77	0.98	64.10	0.67	0.93

The results reported in Tables 2 and 3 indicate that the GDGM provides a clear improvement in the success rate of the classification. Table 2 shows how the performance of conformal predictors degrade with increasing levels of noise. It is important also to notice how the indicators of the quality of the prediction, confidence and credibility, tend to overestimate the reliability of the classification when significant level of noise is present. Table 3 reports the clear improvement in both performance and reliability of the quality indicators when the Euclidean distance is replaced with GDGM. Another important consideration is the fact that, adopting the GDGM metric does not cause any degradation of performance when the data are not affected by noise.

6 Conclusions

In many applications of conformal predictors, the Euclidean distance is explicitly or implicitly adopted as the right metric. In the case of experimental measurements typical of the physical sciences, the data are affected by noise of normal distribution. In this situation, the GDGM proves to be a better metric, to be used in the definition of the non-conformity measure. The calculation of the nonconformity measure and of the p-values using the GDGM provides significantly more reliable classifications, by reducing the adverse effects of the noise. The reported results using the GDGM have been obtained using a desktop computer with two Xeon E5520 @2.27 GHz processors and 24 GB of RAM, and required an average of one minute for each test performed, for a total of 50 min for all 50 points. The computational cost is therefore very similar to the one required to perform the calculations with the Euclidean distance.

With regard to future developments, it would be important to apply the same approach to different pdfs: particularly relevant would be the case of the Poisson distribution, since in practice many detectors work in photon counting or particle counting mode. Another very interesting application would be the case in which the pdf of the noise is not known. This situation has practical applications because in many experimental situations the uncertainties in the measurements can be quantified with an interval but without any additional specification. Therefore the real value is expected to fall in a certain interval but no additional information is available. In this case the implementation of an appropriate form of uncertain probability is expected to produce improvements in the classification of conformal predictors comparable to the case of the GDGM for the case of measurements affected by Gaussian noise.

In terms of practical applications, the mathematics of conformal predictors can be applied to most classifiers, including Fuzzy ones [6]. Therefore the approach can be of extreme help in all the cases, such as disruptions in Tokamaks, where classification is a particularly problematic and difficult task also due to the uncertainties in the mea-surements [7, 8].

References

1. Vovk, V., Gammerman, A., Shafer, G.: Algorithmic Learning in a Random World. Springer, New York (2005)
2. Katz, J.O., Rohlf, F.J.: Function-point cluster analysis. Syst. Zool. **22**(3), 295–301 (1973)
3. Amari, S., Nagaoka, H.: Methods of Information Geometry. Oxford University Press, Oxford (1993)
4. Murari, A., et al.: Nucl. Fusion **53**, 033006, 9 (2013)
5. Vega, J., et al.: Rev Sci. Instrum. **81**, 10E118 (2010)
6. Murari, A., et al.: Ann. Math. Artif. Intell. **74**(1), 155–180 (2015)
7. Murari, A., et al.: Nucl. Fusion **49**, 055028, 11 (2009)
8. Murari, A., Peluso, E., Gelfusa, M., Lungaroni, M., Gaudio, P.: How to handle error bars in symbolic regression for data mining in scientific applications. In: Gammerman, A., Vovk, V., Papadopoulos, H. (eds.) SLDS 2015. LNCS, vol. 9047, pp. 347–355. Springer, Heidelberg (2015). doi:10.1007/978-3-319-17091-6_29

Decision Trees for Instance Transfer

Shuang Zhou[✉], Evgueni Smirnov, Gijs Schoenmakers, and Ralf Peeters

Department of Data Science and Knowledge Engineering, Maastricht University,
P.O. Box 616, 6200 MD Maastricht, The Netherlands
{shuang.zhou,smirnov,gm.schoenmakers,
ralf.peeters}@maastrichtuniversity.nl

Abstract. Instance-transfer learning has emerged as a promising learning framework to boost performance of predictive models for a target domain by exploiting data from source domains. The success of the framework depends on the relevance of the source data to the target data. This paper proposes a decision-tree approach for instance transfer when the source and target data are relevant with respect to a strict subset of input features. Experimental results on real-world data sets demonstrate that the proposed approach outperforms existing instance-transfer approaches when the source and target data are partially related.

1 Introduction

Instance transfer has received significant attention in transfer learning during the last decade [7]. The goal is to improve the predictive models for a *target* domain by exploiting data from a (closely) related *source* domain. The framework is most attractive when the size of target training sample is relatively small and plenty of labeled instances from the source domain are available. The main assumption in instance transfer is that the target domain and the source domain are represented by the same features and share the same class labels, but differ in the underlying distributions [7]. As a result, the main research questions arising in the field of instance transfer are: (1) how to determine the relevance of the source domain to the target domain; and (2) how to select source instances for transfer.

Instance transfer is typically multi-variate [4,5,12,13]. It employs the entire set of features used to represent the data when evaluating the relatedness between target and source domains, and when deciding which source instances to transfer. Although this is a quite reasonable strategy, it may fail when the target and source domains are related only with respect to a strict subset of the features. In this situation, current instance-transfer approaches (e.g. TrAdaBoost [4]) result in either transferring sub-optimal source instances, or training the predictive models on the target sample only.

This paper addresses the aforementioned problem. It proposes a decision-tree approach to instance transfer when the target and source domains are related to each other w.r.t. a subset of the input features. The proposed approach induces decision trees using the standard decision tree algorithm. Instance transfer is realized on the level of feature selection for test nodes of the decision trees.

© Springer International Publishing Switzerland 2016
A. Gammerman et al. (Eds.): COPA 2016, LNAI 9653, pp. 116–127, 2016.
DOI: 10.1007/978-3-319-33395-3_9

More precisely, for each feature the approach first applies the conformal prediction framework [9] to evaluate the relevance of each source instance to the target ones w.r.t that feature, and then selects source instances on a given significance level. Discriminating power of the feature is estimated afterwards on the target instances and the selected source instances using some measures (e.g. Gain Ratio)[1]. Once the estimations of all the features have been done, the approach selects the feature with the maximal discriminating power for the corresponding node of the tree. We note that decision tree induction consists of a series of such feature selection steps. Thus, our decision-tree approach is essentially a multivariate approach to instance transfer that employs an univariate feature-selection operator. This operator does actual instance transfer and does it differently for different features.

Our decision-tree approach is unlikely to suffer from negative instance transfer, which is defined as a situation in which transferring source instances actually degrades the generalization performance of the predictive model for the target domain [7]. The reason is obvious: when the target and source domains are unrelated for any feature there will be no transfer for that feature. However, when the decision tree overfits the training data, the number of target instances associated with a test node tends to be small. In this case the source relevance estimation might become inaccurate. To address this issue we introduce a parameter that controls the minimum number of target instances per test node when instance transfer is allowed.

The remainder of this paper is structured as follows. Section 2 formalizes the task of instance transfer and gives a sketch of decision tree approaches to instance transfer. Sections 3 and 4 serve as introduction to decision trees and the conformal prediction framework respectively. Section 5 presents the proposed instance-transfer decision trees in detail. An experimental analysis is given in Sect. 6. Finally, Sect. 7 concludes the paper.

2   Task Formalization and Solutions

Let X be a k-dimensional space with k features X_i, $i \in \{1, 2...k\}$, and Y be a class set. We consider a target distribution P_T over the labeled space $(X \times Y)$. The target sample T is a set $\{(x_1, y_1), (x_2, y_2), \ldots, (x_m, y_m)\}$ of m independently and identically distributed (i.i.d) instances drawn from P_T. Given a test instance $x_{m+1} \in X$, the target classification task is to find an estimate $\hat{y} \in Y$ for the true class of x_{m+1} according to P_T.

Now consider another distribution over $(X \times Y)$, namely the source distribution P_S. Under the i.i.d assumption we generate a source sample S as a set $\{(x_1, y_1), (x_2, y_2), \ldots, (x_n, y_n)\}$ of n instances drawn from P_S. Assuming that P_T and P_S are different but similar in some sense, we define *the instance-transfer classification task* as a classification task with an auxiliary source sample S in addition to the target sample T. We note that the class of a new (target) test

---

[1] In this context we note that the discriminating-power estimations of different features can be based on different subsets of source instances.

instance is estimated according to the target distribution P_T. This implies that source instances only serve as auxiliary training data for the classification of unseen target instances.

In this paper, we propose two approaches to embed instance transfer into decision trees induction:

- **Prior Selection:** selecting the most relevant source instances prior to decision tree induction using the entire set of features. More specifically, we first select a source subset $S' \subseteq S$ whose members are likely to be generated by the target distribution P_T, and then add S' in the training sample, i.e. $S' \cup T$ is used for training the decision trees. Note that the selected source set is not tailored to any feature.
- **Dynamic Selection:** selecting the most relevant source instances for each feature $X_i, (i = 1, 2...k)$ at any test node N during the induction process. The selected source instances constitute a new subset $S'_i \subseteq S$ specific to the target data T_N (associated with node N) and the feature X_i being considered. In doing so, the discriminating power of feature X_i is estimated based on the combined training sample $S'_i \cup T_N$. Note that the source selection is dynamic in the sense that it selects different source sets for different features and different decision nodes.

Both of the aforementioned approaches are applicable when the target and source domains are related w.r.t the entire set of features. When the target and source domains are related only with respect to a subset of the input features, the prior selection may either transfer sub-optimal source instances, or discard the source instances completely. On the contrary, the dynamic selection allows for transferring instances with respect to a subset of related features. In this manner, overall transfer performance can be improved.

In the paper, we advocate the dynamic selection approach due to its more general applicability. It tailors the selection of source instances to each feature and accommodates the node-specific needs of decision trees.

3   Decision Trees

Decision trees are one of the most widely applied techniques for classification [8]. A decision tree is defined recursively as a leaf node assigning a class or a test node with a decision tree for each test outcome. A decision tree is induced in a top-down manner from a training sample. The algorithm can be written almost entirely as a single recursive method **BuildTree**, as shown in Algorithm 1.

We define three functions to be used in the Algorithm 1:

- **Terminate**(T) is a boolean function that returns true if sample T satisfies a stopping criterion, or false otherwise.
- **DiscriminatingPower**(T, X_j) is a real value function that returns an estimation of the discriminating power of feature X_j on sample T.
- **Split**(T, X_j) is a function that splits sample T into subsamples according to feature X_j.

The decision tree induction starts by calling the **BuildTree** function on the entire training sample T. The function first tests whether the training sample satisfies a stopping criterion (e.g., *entropy* = 0). If so, it creates a leaf node and then stores the number s_y of training instances in T for each class $y \in Y$ that arrive at this node. Otherwise, the function creates a test node N with some test that splits the sample into subsamples for all possible test outcomes. To decide which feature should be tested at node N, each feature is evaluated based on its discriminating power. An ideal split should form subsamples that exhibit class purity, i.e. each subsample should contain instances belonging to only one class. Once evaluations of all features have been obtained, the one with best discriminating power is selected and used as the test at node N. A descendant of the test node N is then created for each possible value of the selected feature, and the training instances are sorted to the appropriate descendant nodes (i.e. down the branch corresponding to the instances's value of this feature). For each of the subsample generated by the split, function **BuildTree** is recursively called. Each call generates a subtree which root is then attached as a child to the principal node N.

Algorithm 1. Decision Tree Induction

Input: Training sample $T \subseteq X \times Y$
Output: Node N

1: **function** BUILDTREE(T)
2:     **if** Terminate(T) **then**
3:         Return a leaf node N with number s_y of instances for each class $y \in Y$.
4:     **else**
5:         Create a test node N.
6:         **for each** feature $X_j \in X$ **do**
7:             DP_j:=DiscriminatingPower(T, X_j).
8:         **end for**
9:         Choose the feature X_j with maximal DP_j as the test feature for node N.
10:        $TargetSubsets := Split(T, X_{test})$.
11:        **for each** $T_i \in TargetSubsets$ **do**
12:            n_i:=BUILDTREE(T_i);
13:            Assign n_i as a child to node N.
14:        **end for**
15:        Return N.
16:     **end if**
17: **end function**

Decision trees classify any instance $x \in X$ in a top-down manner as well: starting at the root of a tree and moving through it until a leaf node is encountered. The scores s_y over all the classes $y \in Y$ associated with the leaf node are normalized to estimate class probability distribution $\{p_y\}_{y \in Y}$ over Y for that instance x.

4   Conformal Prediction Framework

The conformal prediction has been proposed in [11]. It uses past experiences to determine precise levels of confidence in new predictions in one domain (let say the target domain) [10]. Let us consider the target sample T as a sequence of m labeled training instances and let x_{m+1} be a new test instance. A conformal predictor provides an estimation y_{m+1} of the class for x_{m+1} by utilizing a conformity test for the null hypothesis "the sequence $T^* = T \cup \{(x_{m+1}, y_{m+1})\}$ is generated by the target distribution P_T under the exchangeability assumption[2]".

The test is based on nonconformity scores α_i of instances $(x_i, y_i) \in T^*$. The nonconformity score α_i is a value indicating how unusual instance (x_i, y_i) is with respect to all instances in the sequence $T^* \setminus \{(x_i, y_i)\}$. To compute a nonconformity score for an instance, we need an instance nonconformity function A. If $(X \times Y)^{(*)}$ denotes the set of all sequences defined over $(X \times Y)$, then the instance nonconformity function A is a mapping from $(X \times Y)^{(*)} \times (X \times Y)$ to $R^+ \cup \{+\infty\}$ and it indicates how unusual an instance (x_i, y_i) is for the sequence $T^* \setminus \{(x_i, y_i)\}$. We note that any instance nonconformity function has to produce the same result for an instance independently on the permutations of the sequence T^* (otherwise, the instance will have $|T^*|!$ possible nonconformity scores).

The nonconformity score α_{m+1} of the test instance (x_{m+1}, y_{m+1}) is used as a test statistic. Under the null hypothesis, the p-value of the test is calculated as the fraction of instances in T^* that are associated with nonconformity scores that are as extreme as or more than α_{m+1} (as shown in Eq. 1). The larger the p-value is, the more likely to observe this value of the test statistic under the null hypothesis, and the more confidence, therefore, we have in the prediction y_{m+1}.

$$t(T^*, (x_{m+1}, y_{m+1})) = \frac{\#\{i = 1, ..., m+1 : \alpha_i \geq \alpha_{m+1}\}}{m+1} \tag{1}$$

In the context of instance transfer, we consider the source sample S as a sequence consisting of n labeled source instances. We are interested in estimating the relevance of any instance $(x_j, y_j) \in S$ to the target sample. With this purpose, the conformity test can be applied for the null hypothesis "the sequence $T^* = T \cup \{(x_j, y_j)\}$ is generated by the same distribution under the exchangeability assumption". Analogously, the p-value of the test can be computed by Eq. 1. The larger the p-value is, the more likely the source instance (x_j, y_j) is generated by the same distribution as target ones. We regard a source instance as a relevant instance and transfer it if its p-value is equal to or greater than a given significance level ϵ_t.

---

[2] The exchangeability assumption states that the joint probability distributions of a sequence of random variables and any of its permutations coincide. It is weaker than the i.i.d assumption.

5   Instance Transfer with Decision Trees

Our algorithm for instance transfer induces a decision tree using the standard decision tree algorithm (shown in Algorithm 1). The difference lies in the usage of source instances. That is, our algorithm allows for transferring relevant source instances when estimating the discriminating power of features for a test node. In doing so, a function **Transfer DiscriminatingPower** is proposed to replace the **DiscriminatingPower** in Algorithm 1. The proposed function is detailed in Algorithm 2.

Algorithm 2. Transfer DiscriminatingPower

Input: Target sample $T \subseteq X \times Y$, Source sample $S \subseteq X \times Y$,
Feature X_i, Nonconformity function A, Significance level ϵ_t,
Minimum number λ of target instances per class for instance transfer.

Output: Discriminating power of feature X_i

1: **if** the number of instances in T for any class $y \in Y$ is smaller than λ **then**
       return DiscriminatingPower(T, X_i).
2: **else**
3:     Build projection T_i of T in $X_i \times Y$.
4:     Build projection S_i of S in $X_i \times Y$.
5:     Set the relevant source subset $S_i' = \emptyset$.
6:     **for each** source instance $(x_j^i, y_j) \in S_i$ **do**
7:        Set sample T^* equal to $T_i \cup \{(x_j^i, y_j)\}$.
8:        **for each** instance $(x_k, y_k) \in T^*$ **do**
9:           Compute nonconformity score

$$\alpha_k := A(T^* \setminus \{(x_k, y_k)\}, (x_k, y_k)).$$

10:       **end for**
11:       Compute p-value p equal to $t(T^*, (x_j^i, y_j))$
12:       **if** $p \geq \epsilon_t$ **then**
13:          Add (x_j^i, y_j) to S_i'
14:       **end if**
15:     **end for**
       return DiscriminatingPower($T_i \cup S_i', X_i$)
16: **end if**

The function starts with checking if there are sufficient number of target instances per class belonging to node N for instance transfer. In case of not enough target instances available (i.e. the size of any class is smaller than λ), instance transfer is not performed and the discriminating power of feature X_i is estimated based on the target sample only. That is because the estimations of relevance of source instances to the target sample can be inaccurate due to the lack of sufficient target instances. If there are sufficient target instances, a subsample of source instances is selected for feature X_i based on their instance p-values computed by the conformal prediction framework. To be more specific, the function first builds projections T_i and S_i for the target and source samples respectively in a bivariate space $(X_i \times Y)$. Then for each source instance $(x_j^i, y_j) \in S_i$, the function tests if the sequence $T^* = T_i \cup (x_j^i, y_j)$ is generated by the same distribution under the exchangeablity assumption. To that end, the nonconformity scores for every instances in T^* are computed by instance nonconformity function A. The p-value is then computed by the p-value function introduced in Sect. 4. If the p-value is greater than or equal to ϵ_t, instance (x_j^i, y_j) is added to the relevant subset S_i'. Once all relevant source instances are added, the discriminating power of X_i is estimated on $T_i \cup S_i'$ using some standard functions (e.g. Information Gain).

6 Experiment

This section presents our experimental results and initial conclusions. We first introduce the instance-transfer tasks under study in Subsect. 6.1. The experiment setup is provided in Subsect. 6.2. In Subsect. 6.3, we analyze the generalization performance of the proposed instance-transfer decision trees. Subsection 6.4 discusses the influence of parameter λ that controls the minimum number of target instances per class when instance transfer is allowed.

6.1 Instance-Transfer Classification Tasks

In our experiments, we considered five instance-transfer classification tasks defined on real-world data sets that are commonly used in transfer learning research. Each task was given with a target sample and a source sample. Table 1 shows the descriptions of every data sets. The tasks are described below.

- The first instance-transfer task was the landmine detection task [1]. It is a task of detecting landmine in different landmine fields. There are 29 data samples from 29 landmine fields. The 29 samples have different distributions due to various ground surface conditions. For example, Mine 1 to 15 correspond to foliated regions, while Mine 16 to 29 correspond to regions that have bare earth. In our experiment, Mine 29 was used as the target sample and Mine 1 was used as the source sample. To better fit the scenario that target and source samples are related w.r.t a subset of the input features, we manipulated the marginal distribution of the feature with highest gain ratio of the source sample by adding randomly generated noise.
- The second instance-transfer task was to estimate correct/incorrect diagnosis for a patient treated by a medical doctor [13]. The data consists of records of patients treated by two different doctors. Each patient's record was regarded as an instance that represented by 8 bio-markers and a class label indicating whether the corresponding doctor had provided correct/incorrect diagnosis. Those instances were grouped according to the doctors. We used the group with comparably small size as the target sample while the other one as the source sample.

- The third instance-transfer task was the handwritten digital recognition task from the USPS corpus [6]. It is a task of recognizing handwritten digits (0 to 9) automatically scanned from envelopes by the U.S. Postal Service. We applied *Principal component analysis (PCA)* to project the original 16×16 gray scale images to a feature space of 10 dimensions. The digit IDs (0–9) were considered as classes. A target sample and a source sample were randomly drawn from the transformed USPS corpus. Random noise was added into two features corresponding to highest eigenvalues of the source sample, thus the target and source samples were related w.r.t a subset of features.

- The fourth instance-transfer task was wine quality classification task [2]. This task is based on a collection of instances related to red and white variants of Portuguese wine. Each instance is represented by 11 physiochemical features (e.g. PH values) and a grade given by experts between 0 (very bad) and 10 (very excellent). In our experiments, we defined a multi-class classification task based on the grades, i.e. labeling the instances with grade 0 to 4, 5 to 6, and 7 to 10 as "poor", "normal", and "excellent", respectively. A random sample from red wine was used as the target sample and a random sample of white wine was used as the source sample. To better fit our partially related scenario, noise was added to two features with high gain ratios of the source sample.

- The last instance-transfer task was student performance prediction task [3]. It is a task of predicting students' achievements in Mathematics of two Portuguese schools: Gabriel Pereira and Mousinho da Silveira. Each instance is represented by a series of demographic, social, and school related features and a final grade. In our experiments, we defined a binary classification task on the final grades, i.e. pass or not pass the exam. We used instances from school Mousinho da Silveira as the target sample, and instances from school Gabriel Pereira as the source sample.

6.2 Experiment Setup

For all classification tasks, we employed three instance-transfer classifiers: (1) the proposed instance-transfer decision trees that employ dynamic selection of source instances, denoted as DS-DT; (2) instance-transfer decision trees that employ prior selection of source instances, denoted as PS-DT; (3) TrAdaBoost [4], which is one of the most commonly used instance-transfer algorithms. Standard decision trees were employed by TradaBoost as base learners. The iteration number in TrAdaBoost was set to 10. The minimal number λ of target instances per class in DS-DT was set to 10. All decision trees were induced using the well-known decision tree algorithm C4.5 [8]. The discriminating power of features was evaluated in terms of *Gain Ratio*, which is featured in C4.5.

To set up the source instance selection procedure we needed to set up the instance nonconformity function. This setup was done depending on the type of feature. For numerical features, the nearest-neighbor instance nonconformity function [9] was employed. For a target sample T and an instance (x_i, y_i), the nearest neighbor instance nonconformity function A_{NN} outputs a nonconformity

Table 1. Descriptions of the data sets for instance-transfer classification tasks

Task	Number of class	Size $|T|$	$|S|$
Landmine	2	449	690
Medical center	2	81	453
Wine quality	3	159	1499
USPS	10	183	1820
Student performance	2	46	349

score $\frac{\sum_{j=1}^{k} d_{ij}^+}{\sum_{j=1}^{k} d_{ij}^-}$, where k is the number of nearest neighbors, d_{ij}^+ is the j-th shortest distance from x_i to some instances in T having the same class label as x_i, and d_{ij}^- is the j-th shortest distance from x_i to some instances in T having different class labels. For nominal features, the nonconformity score of an instance (x_i, y_i) w.r.t a sample T is computed based on the posterior probability distribution of T as follows: $\frac{\sum_{\bar{y} \in Y \setminus \{y\}} P(\bar{y}|x)}{P(y|x)}$, where $P(y|x)$ is the normalized count of class y given feature x.

The method of evaluation was repeated 10-fold cross validation on the target samples; i.e., the source samples were used as auxiliary ones. The generalization performance of the instance-transfer classifiers was evaluated using Area Under the ROC Curve (AUC). The performance of C4.5 decision trees for the case of no instance transfer was used as baseline. Paired t-test was performed on significance level $\epsilon_e = 0.05$ to find significantly better(worse) results w.r.t the baseline classifier.

6.3   Experimental Results

We experimentally compared the performance of DS-DT, PS-DT and TrAdaBoost on all aforementioned instance-transfer tasks. The results are given in Table 2. The AUCs of DS-DT and PS-DT are given on significance level ϵ_t for source selection ranging from 0 to 1 with step 0.1. Note that there is only one result of TrAdaBoost provided for each task. That is because TrAdaBoost does not employ any significance level as a parameter. In Table 2, significant negative transfer is marked with "-", while performance that is statistically better than the baseline is marked with "*".

As is shown in Table 2, DS-DT gives statistically better results compared to the baseline in most of the experiments and never results in negative transfer. PS-DT gives less significant improvement in all tasks and even results in negative transfer on some significance level ϵ_t for the USPS and student performance tasks. Comparing DS-DT to PS-DT on each significance level ϵ_t, we observe that DS-DT outperforms or performs equally well as PS-DT most of the time. This observation is especially visible for the wine quality and student performance tasks. Among these three classifiers, TrAdaBoost achieves the least appealing

Table 2. AUCs of DS-DT, PS-DT and TrAdaBoost for five real-world tasks. $*(-)$ denotes significantly better(worse) results w.r.t the baseline classifier C 4.5.

Tasks	Base-line	Classifiers	Significance Level ϵ_t										
			1	0.9	0.8	0.7	0.6	0.5	0.4	0.3	0.2	0.1	0
Landmine	0.56	DS-DT	0.57	0.58	0.58	0.58	0.57	0.57	0.60*	0.60*	0.60*	0.57	0.56
		PS-DT	0.56	0.56	0.55	0.57	0.55	0.55	0.60*	0.60*	0.57	0.58	0.54
		TrAdaBoost											0.56
Medical Center	0.59	DS-DT	0.61	0.63	0.62	0.65*	0.66*	0.66*	0.64*	0.66*	0.62	0.61	0.60
		PS-DT	0.59	0.59	0.62	0.59	0.62	0.64*	0.65*	0.61	0.64*	0.61	0.60
		TrAdaBoost											0.59
Wine Quality	0.59	DS-DT	0.62	0.63*	0.66*	0.65*	0.65*	0.64*	0.64*	0.66*	0.63*	0.63*	0.63*
		PS-DT	0.60	0.60	0.61	0.61	0.62	0.63*	0.62	0.62	0.61	0.61	0.60
		TrAdaBoost											0.62
USPS	0.74	DS-DT	0.75	0.74	0.76*	0.78*	0.77*	0.77*	0.77*	0.78*	0.77*	0.78*	0.74
		PS-DT	0.77	0.76	0.76	0.76	0.78*	0.76	0.78*	0.77	0.72	0.72	0.70⁻
		TrAdaBoost											0.74
Student Performance	0.59	DS-DT	0.65*	0.66*	0.65*	0.65*	0.64*	0.68*	0.70*	0.73*	0.70*	0.60	0.75*
		PS-DT	0.59	0.66*	0.62	0.62	0.62	0.62	0.62	0.62	0.53⁻	0.53⁻	0.53⁻
		TrAdaBoost											0.63*

results. It only gives significant improvement for the student performance task, but no improvement at all for other four tasks. Therefore, we conclude that when the target and source samples are related with respect to a strict subset of features, DS-DT is more robust. It does not suffer from negative transfer and achieves better results compared with the other two classifiers.

6.4 Influence of Minimum Size of Target Sample for Transfer

In this subsection, we investigated the effect of the minimum size λ of target sample per class that allows for instance transfer. The main idea of this parameter is to avoid inaccurate estimations of p-values due to a small target sample. Figure 1 presents the performance of DS-DT classifiers employing λ ranging from 10 to 50 with step 10 for the wine quality task. On the x-axis is significance level $\epsilon_t \in [0,1]$ with step 0.1. The plots show the average AUCs of DS-DT classifiers that perform instance transfer at corresponding significance levels.

The plots show that the performance of DS-DT improves as the value of λ increases from 10 to 30. That is because the estimations of relevance of source instances to the target sample become more accurate as the size of target training sample gets bigger. However, the AUCs drop down when λ is raised to 40. The reason is that instance transfer is only allowed at high-level nodes of the tree when λ is set to 40. Therefore, the final model benefits less from instance transfer. When λ increases to 50 or even greater numbers, there is no instance transfer at all.

Moreover, we note that DS-DT suffers less from irrelevant source instances as the value of λ increases. When the significance level ϵ_t is set to a small number (e.g. 0 to 0.2), the performance of DS-DT drops down due to the inclusion of a big amount of irrelevant source instances. Especially when the size of target

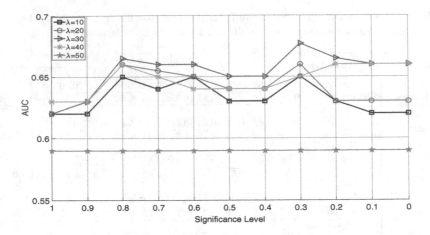

Fig. 1. AUCs of DS-DT classifiers with different λ for the wine quality task.

sample is small, the finally model is skewed to the source distribution. That explains why the DS-DT with lower λ (e.g. $\lambda = 10$ or 20) achieves much worse results than the ones with higher λ at significance level ϵ_t equals 0 to 0.2.

7   Conclusions

Instance transfer has found successful applications in various tasks where the target and source domains are closely related. However, in many real-world scenarios, the target and source domains are partially related, i.e. they are related with respect to part of the features. In this paper we showed that instance transfer can be done with respect to a subset of features. For that purpose we proposed a decision tree approach that employs instance transfer on the level of feature selection at test nodes of the trees. Our approach differentiates itself with the existing approaches through allowing for dynamic and feature-specific source instance selection. Experimental results on five real-world data sets demonstrate that the approach outperforms existing instance-transfer algorithms when the source and target data are related with respect to a subset of features.

References

1. Landmine detection data. http://www.ee.duke.edu/~lcarin/LandmineData.zip
2. Cortez, P., Cerdeira, A., Almeida, F., Matos, T., Reis, J.: Modeling wine preferences by data mining from physicochemical properties. Decis. Support Syst. **47**(4), 547–553 (2009)
3. Cortez, P., Silva, A.M.G.: Using data mining to predict secondary school student performance (2008)
4. Dai, W., Yang, Q., Xue, G.-R., Yu, Y.: Boosting for transfer learning. In: Proceedings of the 24th International Conference on Machine Learning, pp. 193–200. ACM (2007)

5. Kamishima, T., Hamasaki, M., Akaho, S.: TrBagg: a simple transfer learning method and application to personalization in collaborative tagging. In: Proceedings of the 9th IEEE International Conference on Data Mining, pp. 219–228 (2009)
6. Lichman, M.: UCI machine learning repository (2013)
7. Pan, S.J., Yang, Q.: A survey on transfer learning. IEEE Trans. Knowl. Data Eng. **22**(10), 1345–1359 (2010)
8. Ross Quinlan, J.: C4.5: Programs for Machine Learning. Morgan Kaufmann Publishers Inc., San Francisco (1993)
9. Shafer, G., Vovk, V.: A tutorial on conformal prediction. J. Mach. Learn. Res. **9**, 371–421 (2008)
10. Vovk, V.: The basic conformal prediction framework. In: Conformal Prediction for Reliable Machine Learning Theory, Adaptations and Applications, pp. 1–20 (2014)
11. Vovk, V., Gammerman, A., Shafer, G.: Algorithmic Learning in A Random World. Springer, US (2005)
12. Yao, Y., Doretto, G.: Boosting for transfer learning with multiple sources. In: Proccedings of the 21 IEEE Conference on Computer Vision and Pattern Recognition, (CVPR), pp. 1855–1862. IEEE (2010)
13. Zhou, S., Smirnov, E.N., Peeters, R.: Conformal region classification with instance-transfer boosting. Int. J. Artif. Intell. Tools **24**(6), 1560002 (2015)

Hidden Markov Models with Confidence

Giovanni Cherubin[1,2]([✉]) and Ilia Nouretdinov[2]

[1] Information Security Group, Egham, UK
Giovanni.Cherubin.2013@live.rhul.ac.uk
[2] Computer Science Department and Computer Learning Research Centre,
Royal Holloway University of London, Egham Hill, Egham, Surrey TW20 OEX, UK
ilia@cs.rhul.ac.uk

Abstract. We consider the problem of training a Hidden Markov Model (HMM) from fully observable data and predicting the hidden states of an observed sequence. Our attention is focused to applications that require a list of potential sequences as a prediction. We propose a novel method based on Conformal Prediction (CP) that, for an arbitrary confidence level $1 - \varepsilon$, produces a list of candidate sequences that contains the correct sequence of hidden states with probability at least $1 - \varepsilon$. We present experimental results that confirm this holds in practice. We compare our method with the standard approach (i.e.: the use of Maximum Likelihood and the List–Viterbi algorithm), which suffers from violations to the assumed distribution. We discuss advantages and limitations of our method, and suggest future directions.

Keywords: Conformal Prediction · Hidden Markov Models · List–Viterbi algorithm

1 Introduction

Hidden Markov Models (HMMs) are statistical models that have had a great impact in numerous fields since their introduction. They have been widely applied to diverse fields, ranging from Cryptanalysis to Speech Analysis, and they are the state-of-the-art in many applications such as Speech Recognition [4].

The idea behind HMMs is that there exists a time evolving "hidden" process, which we cannot directly observe, and an observable random variable, whose values are related in probability to those of the hidden process. HMMs can be discrete, if the observed process can only take a finite number of values, or continuous, if it takes values from an infinite set. This paper will focus on continuous HMMs. The following problems are of fundamental interest to real-world applications of HMMs: (i) what is the probability that a sequence of observations was generated by an HMM (*evaluation*); (ii) what is the hidden sequence that produced a sequence of observations (*decoding*); (iii) how can we estimate the parameters for an HMM from empirical observations (*learning*).

This paper considers the learning and decoding problems when fully observable data is available and a list of sequences is required as a prediction. That

© Springer International Publishing Switzerland 2016
A. Gammerman et al. (Eds.): COPA 2016, LNAI 9653, pp. 128–144, 2016.
DOI: 10.1007/978-3-319-33395-3_10

is, it assumes a training set that contains data from both the hidden and the observable processes, and it aims at producing, for a new observed sequence, a list of candidate hidden sequences.

The standard approach to this problem is to assume a distribution for the emission probabilities of the HMM, to estimate the parameters of the model by using Maximum Likelihood, and to use the List–Viterbi algorithm [5] to produce a list of candidate sequences. However, the standard approach: (i) requires to manually trim the size of the list in order to achieve the desired level of accuracy, and (ii) can have bad performances if the data does not follow the assumed probability distribution.

We propose a novel approach that: (i) guarantees the accuracy is as good as, or better than, a chosen confidence level, and (ii) makes no assumptions on the probability distribution of the examples, as long as they are exchangeable. The method works in two phases. In the first phase, it uses Conformal Prediction (CP) [9] to replace the estimation of emission probabilities. It accepts a significance level ε as a parameter, and produces a list of candidate hidden sequences that is guaranteed to contain the correct sequence with probability of at least $1 - \varepsilon$ (*validity guarantee*). In the second phase, it ranks the candidate sequences by their likelihood, using estimates of the initial and transmission probabilities. The method returns the list of candidate hidden sequences sorted with respect to their rank. While this paper focuses on continuous HMMs, this method can work on both discrete and continuous HMMs.

Originally, CP worked under the assumption of exchangeability, a weaker property than i.i.d., on training and test data. CP performs well, and gives valid confident prediction under this assumption. However, applying HMM goes beyond exchangeability. The book [9] suggests On-line Compression Models as an extension for various other assumptions including Markov Model (Chap. 8.6). However, this is not directly applicable to HMMs.

We perform experiments to verify the validity guarantee of the method. We also provide a comparison with the standard method. Experiments are made: (i) under optimal conditions for the standard method (i.e.: the data reflect the assumptions it made), (ii) violating the distribution assumed by the standard method. Results show that, while the standard method gives a better accuracy when the assumed emission probability distribution is correct, its performances strongly suffer when this assumption is violated. The method we propose does not depend on the underlying distribution, and provides the desired accuracy level under different distributions of data. Furthermore, it is able to keep the size of the prediction set small under both conditions (*efficiency* criterion).

We conclude our analysis discussing advantages and limitations of the method and suggesting future research directions.

2   Hidden Markov Models

We consider a discrete–time Markov chain q_t, with finite state space. That is, q_t is a random process that at time $t = 1, 2, \ldots$ takes values in a finite set of states S, and for which holds the Markov property:

$$P(q_t = s_t | q_{t-1} = s_{t-1}, q_{t-2} = s_{t-2}, ..., q_1 = s_1) = P(q_t = s_t | q_{t-1} = s_{t-1}),$$

for $s_i \in S$; informally, this property means that the transition of q_t from one state to the next one only depends on its current state.

In a Hidden Markov Model (HMM) there exists a "hidden" Markov chain q_t, as the one we described, whose values are generally unobservable. Whilst we cannot directly observe q_t, we have access to a random variable v_t, whose value at time t depends in probability on the state of q_t. The variable v_t takes values in a measurable space O. In a discrete HMM O is finite, in a continuous one it is infinite. This paper will focus on the continuous case. Figure 1 shows the structure of an HMM.

A continuous *HMM* is defined by a transition probability matrix A, emission probability densities B, and initial probabilities Π. Follows a description of them. A transition probability matrix is a matrix $A = \{\alpha_{ij}\}$, where α_{ij} is the probability that the hidden process makes a transition from state s_i to state s_j:

$$\alpha_{ij} = P(q_t = s_j | q_{t-1} = s_i).$$

We assume that, for each hidden state $s_j \in S$, the conditional distribution:

$$P(v_t | q_t = s_j)$$

has a density function b_j on O. $B = \{b_j\}$, for all $s_j \in S$, is the set of emission probability densities. We also define the initial probabilities $\Pi = \{\pi_i\}$, where:

$$\pi_i = P(q_1 = s_i).$$

We call *observations* the values $o_t \in O$ taken by the observable random variable v_t. We refer to a sequence of contiguous observations as

$$x = (o_1, o_2, ...),$$

where $o_t \in O$ is the value taken by v_t at time t. Analogously, we write

$$h = (s_1, s_2, ...),$$

to indicate a sequence of hidden states. We use the notation $x^{(j)}$ when referring to the j-th element of a sequence x; for example, $x^{(j)} = o_j$ for the sequence

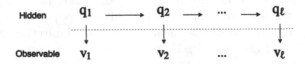

Fig. 1. Structure of an HMM, observed at time $t = 1, 2, ..., \ell$. A Markov chain q_t is hidden, and makes transitions between states $s_i \in S$ with respect to a transition probability matrix A. We can observe a random variable v_t, whose values $o_i \in O$ depend in probability on the current state of q_t; B defines the emission probabilities from a state to the observation.

mentioned above. Similarly, $h^{(j)}$ is the j-th element of the sequence h. In the formulation of the problem (Sect. 3) we will assume that we can fully observe an HMM for ℓ time during a training phase. This operation produces an observable sequence $x = (o_1, o_2, ..., o_\ell)$, and a hidden sequence $h = (s_1, s_2, ..., s_\ell)$.

3 Problem Setting and Evaluation Criteria

We assume we can fully observe an HMM in a *training phase*. In this phase we collect a multiset of n pairs:

$$\{(x_i, h_i)\},$$

where $x_i = (o_1, o_2, ..., o_{\ell_i})$, $o_t \in O$, is a sequence of observations, and $h_i = (s_1, s_2, ..., s_{\ell_i})$, $s_t \in S$, is the respective sequence of hidden states. We assume $|x_i| > 1$, for $i = 1, 2, ..., n$, but we do not require that $|x_i| = |x_j|$ for $i \neq j$.

In a *test phase* we are given a new sequence of observations x_{n+1}, whose corresponding hidden sequence h_{n+1} is unknown to us. Our goal is to predict a list of candidate hidden sequences $\hat{H}$, sorted by their likelihood, that contains the correct hidden sequence.

We consider three evaluation criteria for the problem:

Accuracy: an error is made when the correct sequence is not in the prediction set $\hat{H}$. Let η be the number of errors committed in n predictions, accuracy is:

$$1 - \frac{\eta}{n}.$$

Efficiency: is the average size of the prediction set (see N *criterion* in [8]). This criterion is crucial to the problem: a perfect accuracy can be achieved by trivially returning the list of all the possible sequences of length $\ell = |x_{n+1}|$; however, it is more difficult to achieve a good accuracy while keeping small the size of $|\hat{H}|$.

Average Position (AP): this criterion evaluates the goodness of the ranking scores we associate with the predicted sequences. AP is the average position that the correct sequence takes within the sorted prediction list $\hat{H}$.

4 Standard Approach

The standard approach to the problem is as follows: a family of probability distributions is assumed for emissions; the parameters of these distributions and initial and transition probabilities are estimated from training data by using Maximum Likelihood; then, the List–Viterbi algorithm is applied, for a certain value of k, to predict the sequence of hidden states. The List–Viterbi algorithm returns a list of k candidate sequences. If the application requires some confidence that the correct sequence is in the predicted list, experiments need to be done to determine which value of k gives the desired accuracy.

This section presents the Maximum Likelihood method to estimate the parameters of an HMM from fully observed data (observations and hidden states), and the List–Viterbi algorithm [5], an extension of the Viterbi algorithm [1, 7], which outputs the k best sequences.

4.1 Maximum Likelihood Method for Estimating A, B, Π

Let $Z = \{(x_i, h_i)\}$, for $i = 1, 2, ..., n$, be a multiset of observed sequences x_i and corresponding hidden sequences h_i. We shall use this multiset for estimating A, B, Π. Let S be the set of hidden states, and N its size.

Initial Probabilities. Initial probabilities Π can be estimated as follows:

$$\Pi = \{\pi_j\} = \left\{ \frac{|\{i : h_i^{(1)} = s_j \quad (x_i, h_i) \in Z\}|}{n} \right\} \quad j = 1, 2, ..., N.$$

Transition Probabilities. Let Z' be a multiset of pairs composed of the hidden state at time t and the hidden state at time $t + 1$. We derive Z' as:

$$Z' = \{(x_i^{(t)}, x_i^{(t+1)})\} \quad t = 1, 2, ..., (\ell_i - 1) \quad (x_i, h_i) \in Z,$$

where $\ell_i = |x_i|$. The probability of transitioning from s_i to s_j is estimated as:

$$\alpha_{ij} = \frac{|\{(s_t, s_{t+1}) \in Z' : s_t = s_i \wedge s_{t+1} = s_j\}|}{|\{(s_t, s_{t+1}) \in Z' : s_t = s_i\}|},$$

and is done for all $i = 1, 2, ..., N$ and $j = 1, 2, ..., N$. The transition probability matrix is $A = \{\alpha_{ij}\}$.

Emission Probabilities. Estimation of emission probability densities $B = \{b_j\}$, for $s_j \in S$ depends on the chosen probability density. A typical choice is the Normal density function: $b_j \sim \mathcal{N}(\mu_j, \sigma_j)$, for some mean μ_j and standard deviation σ_j.

Let Z'' be a multiset of pairs composed of an observable state and the corresponding hidden state:

$$Z'' = \{(x_i^{(j)}, h_i^{(j)})\} \quad j = 1, 2, ..., \ell_i \quad (x_i, h_i) \in Z,$$

where $\ell_i = |x_i|$. We estimate the parameters for b_j (e.g.: μ_j, σ_j for a Normal density) on the multiset:

$$\{o : (o, s) \in Z'' \wedge s = s_j\}.$$

4.2 Viterbi Algorithm

The Viterbi algorithm computes the most likely sequence of states $\hat{h}$ for an observed sequence $x = (o_1, o_2, ..., o_\ell)$, given an HMM (A, B, Π).

At each step $t = 1, 2, ..., \ell$ the Viterbi algorithm computes, for each state $s_i \in S$, the probability $V_t(s_i)$ of the most likely sequence for which $q_t = s_i$. It first initialises:

$$V_1(s_i) = P(o_1 | q_1 = s_i) P(q_1 = s_i) \quad s_i \in S,$$

where $P(o_1|q_1 = s_i) = b_{s_i}(o_1)$, and $P(q_1 = s_i) = \pi_i$. Then, for each step $t > 1$, it sets the probability of being at state s_i at time t, $V_t(s_i)$ to:

$$V_t(s_i) = P(o_t|q_t = s_i) \max_{s_j \in S} P(q_t = s_i|q_{t-1} = s_j)V_{t-1}(s_j),$$

for all $s_i \in S$. We remark that $P(o_t|q_t = s_i) = b_{s_i}(o_t)$ and $P(q_t = s_i|q_{t-1} = s_j) = \alpha_{ji}$. $V_t(s_i)$ represents the probability of being in state s_i at time t, given that the most likely path to reach s_i was followed by the HMM.

The most likely sequence can be obtained by using back pointers to the best path taking to each state, for time $t = 1, 2, ..., \ell$.

4.3   List–Viterbi Algorithm

The List–Viterbi algorithm is an extension of the Viterbi algorithm, which outputs the k most likely hidden sequences for the observed sequence x.

The algorithm works as the Viterbi algorithm, but each variable $V_t(s_i)$, for $t > 1$, is a vector of length k; the j-th element of vector $V_t(s_i)$ is the likelihood of the j-th most likely sequence that takes to state s_i at time t. At each step $t > 2$, all the $k|S|$ likelihoods are considered, and only the best k are kept for the next step. The List–Viterbi algorithm returns a list of the most likely sequences, which are obtained by using back pointers to the k best paths. The sequences of the prediction list are sorted by their likelihood.

5   Prediction with Confidence for HMMs

This section introduces a method to train an HMM from fully observable data and to make a prediction for a new observed sequence. The method outputs a list of candidate hidden sequences $\hat{H}$, sorted with respect to their likelihood; $\hat{H}$ contains the correct sequence with probability at least $1 - \varepsilon$, for a chosen significance level ε.

The method operates in two phases. In the first phase, the algorithm:

1. uses training data to create a training set Z_{train} of pairs (o_i, s_i), for observations $o_i \in O$ and respective hidden states $s_i \in S$;
2. considers each observation o_j of the test sequence individually, and uses CP and the training set Z_{train} to determine a set of candidate hidden states $\hat{H}_j$ for that observation; when doing this, hidden states are considered as the labels to predict;
3. produces the list $\hat{H}$ of all the hidden sequences that can be generated by using one state from $\hat{H}_1$ as a first state, one from $\hat{H}_2$ as a second state, and so on;

Figure 2 offers a graphical overview of the first phase.

The second phase is concerned with sorting the list of candidate hidden sequences $\hat{H}$ by their likelihood. In this phase the algorithm computes Maximum Likelihood estimates of initial and transition probabilities. It computes a ranking score for each sequence, using the Maximum Likelihood estimates, as the probability of the hidden Markov chain q_t to produce that sequence. The algorithm returns a list $\hat{H}$ of sequences, sorted with respect to their ranking scores.

We introduce CP, and present the method into details.

5.1   Conformal Prediction

CP is a statistical framework that allows to edge predictions with respect to a confidence level [2,6,9]. Let $z_i = (o_i, s_i)$, for $i = 1, 2, ..., n$, be pairs of observation and respective hidden state, and $\varepsilon \in [0, 1]$ a significance level. We identify a hidden state with the label to predict. CP produces, for a new observation o_{n+1}, a set of candidate labels Γ^ε. The validity property of CP guarantees that Γ^ε contains the correct label, s_{n+1}, with probability $1 - \varepsilon$, for an arbitrary significance level[1]. We call $1 - \varepsilon$ confidence level.

Nonconformity Measure. CP works for a *nonconformity measure*:

$$A : O^{(*)} \times O \mapsto \mathbb{R}.$$

The function A accepts a multiset of observations and a new observation, and returns a scalar (nonconformity score) that indicates how strange the new observation is respect to the multiset. Any function in the form of A guarantees the validity of the method. However, some functions may provide a better efficiency in the terms described in Sect. 3.

In our analysis, we consider the k-Nearest Neighbours (k-NN) nonconformity measure, which is computed as follows. Let $\mathcal{O}$ be a multiset of observations, o_{n+1} a new observation, and δ_i the i-th smallest distance between o_{n+1} and the observations in $\mathcal{O}$. The k-NN nonconformity measure is:

$$A(\mathcal{O}, o_{n+1}) = \sum_{j=1}^{k} \delta_i,$$

where k is the chosen number of neighbours. In experiments we will use the k-NN nonconformity measure with $k = 1$.

CP in Multi–label Setting. Different formulations of CP exist. We consider the multi–label setting, where we are given examples (o_i, s_i) of observation $o_i \in O$ and label $s_i \in S$, and CP returns, for a new observation, a set of candidate labels $\Gamma^\varepsilon \subseteq S$.

Algorithm 1 describes CP in this setting. We write:

$$\Gamma^\varepsilon = CP(o_{n+1}, Z, A, \varepsilon)$$

---

[1] This paper will write CP implicitly indicating Smooth CP. The difference is that standard CP would guarantee ε to be an upper bound of errors [9].

to indicate a call to this algorithm for a new observation o_{n+1}, a training set Z, nonconformity measure A, and significance level ε. Thanks to the validity property of CP, Γ^{ε} is guaranteed to contain the correct label for o_{n+1} with probability $1 - \varepsilon$.

Algorithm 1. Smooth Conformal Prediction in multi–label setting.

Require: Multiset of examples $Z = \{z_1, z_2, ..., z_k\}$, where each example z_i is a pair (o_i, s_i) of an observation $o_i \in O$ and a label $s_i \in S$, nonconformity measure A, significance level ε, new observation o_{k+1}.

  Create empty list Γ^{ε}.
  for $\hat{s}$ in S **do**
    Set provisionally $Z = \{z_1, .., z_k, (o_{k+1}, \hat{s})\}$
    $O_{\hat{s}} \leftarrow \{o_i | (o_i, s_i) \in Z, s_i = \hat{s}\}$
    for $i \leftarrow 1, 2, ..., k + 1$ **do**
      $\alpha_i \leftarrow A(O_{\hat{s}} \setminus o_i, o_i)$
    end for
    $\tau \leftarrow Uni(0, 1)$ $\triangleright$ Sample τ from the uniform distribution in $[0, 1]$.
    $p_k \leftarrow \frac{|\{i : \alpha_i > \alpha_k\}| + |\{i : \alpha_i = \alpha_k\}| \tau}{k}$
    if $p_k > \varepsilon$ **then**
      Add $\hat{s}$ to list Γ^{ε}
    end if
  end for
  return Γ^{ε}

5.2 Prediction with Confidence for HMMs

We are provided with a multiset of pairs $Z = \{(x_i, h_i)\}$, for $i = 1, 2, ..., n$, of observable and respective hidden sequences (Sect. 3). We are also given a test sequence $x_{n+1} = (o_1, o_2, ..., o_\ell)$, whose corresponding hidden sequence h_{n+1} is unknown to us. Follows a description of the method for making a prediction with confidence for h_{n+1}.

The method is composed of two phases, that we shall call *Confident Prediction* and *Ranking*. The former aims at producing a list of candidate sequences $\hat{H}$ that contains h_{n+1}. The latter computes the likelihoods (ranking scores) of the sequences in $\hat{H}$, and returns the list sorted with respect to them.

Confident Prediction. The first phase uses information about the relation between hidden states and observations to make a list prediction for a new sequence.

We create a multiset of pairs of observation and respective hidden state:

$$Z_{train} = \{(x_i^{(k)}, h_i^{(k)})\} \quad k = 1, 2, ..., \ell_i \quad (x_i, h_i) \in Z,$$

where ℓ_i is the length of the i-th sequence. We will consider Z_{train} as a training set, where hidden states are the labels to predict from observations.

We individually consider each observation $o_j = x_{n+1}^{(j)}$ from the sequence x_{n+1}, for $j = 1, 2, ..., \ell$, and look for candidate hidden states for it. Specifically, we use CP and the training set Z_{train} to predict a set of labels (hidden states) $\hat{H}_j$ for the observation o_j:

$$\hat{H}_j = CP\left(o_j, Z_{train}, A, \frac{\varepsilon}{\ell}\right),$$

where CP is Smooth CP in multi–label setting (Algorithm 1). Any nonconformity measure A in the form described in Sect. 5.1 is allowed, but some nonconformity measures may provide a better efficiency. The result of this operation is a set $\hat{H}_j$ containing candidate hidden states for the observation o_j. We assume exchangeability on the elements of the multiset:

$$Z_{train} \cup (x_{n+1}^{(j)}, h_{n+1}^{(j)}).$$

Then, thanks to the validity property of CP, $\hat{H}_j$ contains the correct hidden state $h_{n+1}^{(j)}$ with probability $1 - \frac{\varepsilon}{\ell}$.

We iterate this operation for each observation o_j of the sequence $x_{n+1} = (o_1, o_2, ..., o_\ell)$, obtaining the sets $\hat{H}_1, \hat{H}_2, ..., \hat{H}_\ell$. We obtain ℓ sets of candidate hidden states, each one indicating candidate states for a position in the sequence.

We produce all the sequences of length ℓ having as a first state one state from $\hat{H}_1$, as a second state one from $\hat{H}_2$, and so on. This means we take the Cartesian product of these sets:

$$\hat{H} = \hat{H}_1 \times \hat{H}_2 \times ... \times \hat{H}_\ell.$$

We call $\hat{H}$ the prediction list. The probability that h_{n+1} is in $\hat{H}$ is:

$$P(h_{n+1} \in \hat{H}) \geq 1 - \varepsilon,$$

for an arbitrary significance level $\varepsilon \in [0, 1]$. A proof of this is given in Appendix.

Ranking. The second phase of the algorithm focuses on ranking the sequences of $\hat{H}$ with respect to their likelihood.

We estimate initial and transition probabilities (A, Π) using Maximum Likelihood (Sect. 4.1). Then we compute a ranking score $\sigma(\hat{h})$, for the hidden sequence $\hat{h} \in \hat{H}$, $\hat{h} = (s_1, s_2, ..., s_\ell)$, as the probability that the hidden process of the HMM produced that sequence:

$$\sigma(\hat{h}) = P(\hat{h}|\Pi, A) = P(s_1) \cdot \prod_{t=1}^{\ell-1} P(s_t|s_{t-1})$$

$$= \pi_{s_1} \cdot \prod_{t=1}^{\ell-1} \alpha_{s_{t-1}s_t};$$

here π_{s_1} is the initial probability for state s_1, $\alpha_{s_{t-1}s_t}$ is the probability of transitioning from state s_{t-1} to state s_t.

We return the list $\hat{H}$ sorted with respect to the ranking scores of its sequences. A larger score gives a higher position in the list.

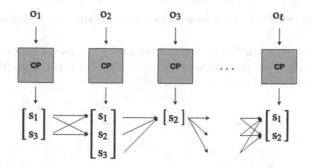

Fig. 2. The first phase of prediction with confidence for HMMs. A test sequence $o_1, o_2, ..., o_\ell$ is observed. We apply CP individually to each observation o_j using a training multiset of observations and respective hidden states. This returns, for each o_j of the sequence, a list of candidate hidden states $\hat{H}_j$. We produce the list $\hat{H}$ of all the sequences that can be generated by using one of $\hat{H}_1 = \{s_1, s_2\}$ as the first state, one of $\hat{H}_2 = \{s_1, s_2, s_3\}$ as the second state, and so on. In the second phase the sequences are ranked with respect to their initial and transition probability estimates.

6 Experiments

In this section we show that the validity property of our method holds in practice. This means that, for different significance levels ε, the method keeps an error which is always smaller or equal to ε. Furthermore, we present an experimental comparison of our method with the standard approach. Similarly to what [3] did when comparing the Bayes approach and CP, we experiment with these methods under two settings: (i) emission probabilities follow the distribution assumed by the standard approach (optimality for the standard approach), (ii) emissions violate this distribution. This approach needs generating two datasets that fulfill these requirements. We refer to these datasets as *HMM-NORM* and *HMM-GMM*. *HMM-NORM* was generated by a continuous HMM, for which emission probabilities were normally distributed. *HMM-GMM* was generated by a continuous HMM, which used mixtures of Normal distributions (GMM) as emission probability densities. Construction details are in Appendix.

In experiments, we consider an on–line setting, where the correct sequence is provided after prediction, and the predicted example is added to the training set. Our training set starts from 4 observed sequences and reaches 2000.

6.1 Validity of the Method

The method we propose is valid, in the sense that it produces a prediction set that contains the correct sequence with probability at least $1 - \varepsilon$, for an arbitrary significance level ε. A proof of this is Appendix.

We apply our method to the data for significance levels: $(0.01, 0.05, 0.1)$, and nonconformity measure k-NN, for $k = 1$. Figure 3 shows the cumulative error of the method in this setting. We observe that the validity property holds empirically: the error tends to be equal or smaller than the significance levels, for a chosen level.

Figure 4 compares the significance level and the respective empirical error that was achieved. This plot shows that the empirical error is smaller than the significance level for each value.

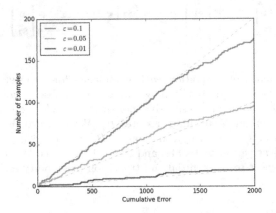

Fig. 3. Cumulative error of our method on the *HMM-NORM* dataset. The validity is respected empirically for each significance level. We refer to our method as *CP-HMM*.

6.2   Comparison with the Standard Approach

We compare our method with the standard approach on datasets *HMM-NORM* and *HMM-GMM*. We assume Normal distribution for the emission probabilities of the standard approach. Consequently, *HMM-NORM* represents the optimal conditions for the standard approach. *HMM-GMM* violates its assumptions.

Accuracy for the Same Size of Prediction Set. We measure the accuracy of our method and of the standard approach when producing a set of predictions of the same size. In order to do this we first run our method for some significance level ($\varepsilon = 0.01$), we record the size of the prediction list $\hat{H}$, and we run the List–Viterbi algorithm for $k = |\hat{H}|$. Results of this experiment on *HMM-NORM* and *HMM-GMM* are shown in Fig. 5.

We observe that the standard approach achieves the best accuracy under optimal conditions (Fig. 5(a)). In this case, our method achieves a slightly worse accuracy than the standard approach. However, when the assumptions of the standard approach are violated (i.e.: emission probabilities are not normally distributed), its error increases considerably (Fig. 5(b)). Nonetheless, our method is able to keep the same accuracy as before (see Fig. 5(b)). This suggests that our method may be applied to a wider range of cases, where estimating the probability distribution of emissions is non–trivial.

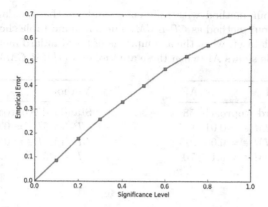

Fig. 4. Average error achieved by our method, for different significance levels. The empirical error tends to be smaller than ε. We refer to our method as *CP-HMM*.

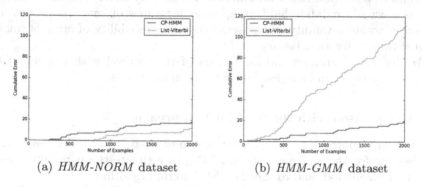

(a) *HMM-NORM* dataset   (b) *HMM-GMM* dataset

Fig. 5. Cumulative error of our method (which we call, for brevity, *CP-HMM*) and the standard approach, when they produce a prediction set of the same size. The left figure shows results under optimal conditions for the standard approach. The right figure shows what happens when its assumptions are violated.

Average Position. In this experiment we determined the Average Position (AP) of our method and of the standard method. Namely, we determined which of the two methods puts the correct prediction closer to the top of their prediction lists. This criterion helps to understand what is the smallest size of the prediction list that achieves perfect accuracy. A smaller AP indicates a better performance.

Table 1 reports the average position taken by the correct prediction in the prediction list, when using the List–Viterbi algorithm, and confidence prediction for HMMs (for significance levels $(0.01, 0.05, 0.1)$).

We notice that AP of our method tends to get better for higher significance levels. The standard approach under its optimal conditions is better, in terms of AP. However, we observe that its AP gets much worse when the data violates its assumptions (*HMM-GMM*). In this case our method is able to perform better.

Table 1. AP for our method with different ε and for the standard approach. For brevity, we refer to our method as *CP-HMM*. The left–hand table shows the results for the *HMM-NORM* dataset, when the assumptions of the standard method are satisfied; the right–hand table shows AP when these are violated (*HMM-GMM* dataset).

Method	AP
Standard Approach	58
CP-HMM $\varepsilon = 0.01$	917
CP-HMM $\varepsilon = 0.05$	208
CP-HMM $\varepsilon = 0.1$	70

Method	AP
Standard Approach	294
CP-HMM $\varepsilon = 0.01$	1067
CP-HMM $\varepsilon = 0.05$	337
CP-HMM $\varepsilon = 0.1$	146

7 Conclusions

We proposed a method that trains an HMM from fully observable data and that outputs a list of candidate hidden sequences for a new observed sequence. The method guarantees validity, in the sense that its probability of error is smaller or equal than ε, for an arbitrary $\varepsilon \in [0, 1]$.

We discuss advantages and limitations of the method with respect to the standard approach, and suggest future research directions.

7.1 Comparison with the Standard Approach

The standard approach to the problem we considered is to assume probability distributions for the emissions of the HMM, to estimate the parameters using Maximum Likelihood, and to use the List–Viterbi algorithm.

One limitation of the List–Viterbi algorithm is that it does not allow to directly control the accuracy. We thus need trim on experimental data the parameter k, that indicates the size of the prediction list, and choose the value that gives the desired level of accuracy. The method we propose accepts a significance level ε, and guarantees that its error is upper–bounded by ε. This means that our method gives a direct control over the errors.

The standard approach is optimal when the correct distributions are assumed, and the parameters are correctly estimated. However, if the data assume different probability distributions, its performances strongly deteriorate. Results on both optimal and non–optimal conditions for the standard method show that the method we propose is robust independently of the distributions. For this reason, we suggest that our method may have a wider applicability to complex cases, where estimating the correct distributions is non–trivial.

As an advantage with respect to the standard method, our method reduces the state space (first phase of the method, Sect. 5.2). While the standard method needs to consider any state as a candidate, given an observation, our method allows to consider only those that conform the distribution. Future work may try to apply variants of the Viterbi and List–Viterbi algorithms to the result of the first phase of our method, as a way of reducing their complexity.

One disadvantage of our method is that CP might return an empty set as a prediction for an observation. This would cause an empty prediction list. To overcome this problem, we may modify Algorithm 1 to output some states, even when none of them conforms. Future research may experiment with this option, and perhaps verify if this would affect the validity of the method.

7.2   Future Work

Future work may apply our method to real–world problems. The method is applicable to both discrete and continuous HMMs, and it has the advantages of: (i) being independent of the probability distributions, and (ii) providing a direct control on the errors.

Our experiments focused on the k-NN nonconformity measure, but the method can work for any nonconformity measure (Sect. 5.1). However, as for CP, some nonconformity measures may provide tighter predictions. Future research may consider other nonconformity measures, such as Kernel Density Estimation, and determine if they can achieve better performances.

Our method, in its current form, uses information about transition and emission probabilities in two separate phases. CP is used in the emission phase only. Although the method made the prediction better, the following challenge appears for the future. If an observation of the hidden sequence does not look to come from its true hidden state (e.g.: there is noise between the hidden process and the random variable), the method will not consider further information (e.g.: transition probabilities) when making a prediction. Future research may attempt to solve this problem. One way is using probabilistic Venn–Machines [9], which may substitute CP in our method. One advantage of them would be a probabilistic output, which may be combined with initial and transition probabilities.

Future work may also consider other ways to rank the predicted sequences, in order to improve the Average Position of the method. The use of Venn–Machines may be helpful also in this case.

Finally, future research may try to limit the size of the training data to reduce the complexity of the method (e.g.: Inductive CP).

Acknowledgements. Giovanni Cherubin was supported by the EPSRC and the UK government as part of the Centre for Doctoral Training in Cyber Security at Royal Holloway, University of London (EP/K035584/1). This project has received funding from the European Unions Horizon 2020 Research and Innovation programme under Grant Agreement no. 671555 (ExCAPE). This work was also supported by EPSRC grant EP/K033344/1 ("Mining the Network Behaviour of Bots"); by Thales grant ("Development of automated methods for detection of anomalous behaviour"); by the National Natural Science Foundation of China (No.61128003) grant; and by the grant "Development of New Venn Prediction Methods for Osteoporosis Risk Assessment" from the Cyprus Research Promotion Foundation.

We are grateful to Alexander Gammerman, Kenneth Paterson, and Vladimir Vovk for useful discussions. We also would like to thank the anonymous reviewers for their insightful comments.

A Validity of the Method

We are given a multiset (training set) of sequences $\{(x_i, h_i)\}$, for $i = 1, 2, ..., n$. We select a significance level $\varepsilon \in [0, 1]$. Let x_{n+1} be a test sequence and h_{n+1} the corresponding sequence of hidden states. Our method outputs a prediction set $\hat{H} = \{h_1, h_2, ...\}$. We show that the probability that $\hat{H}$ contains the correct sequence is at least $1 - \varepsilon$.

Let us construct the following multiset:

$$Z_{train} = \{(x_i^{(j)}, h_i^{(j)})\} \quad j = 1, 2, ...\ell_i \quad i = 1, 2, ..., n,$$

where $\ell_i = |x_i| = |h_i|$.

Let $\ell = |x_{n+1}| = |h_{n+1}|$. Let us consider the j-th element of the sequence x_{n+1}. We assume exchangeability on the multiset

$$Z_{train} \cup \{(x_{n+1}^{(j)}, h_{n+1}^{(j)})\}.$$

We run:

$$\hat{H}_j = CP\left(x_{n+1}^{(j)}, Z_{train}, A, \frac{\varepsilon}{\ell}\right),$$

as defined in Algorithm 1. Thanks to the validity property of Smooth CP [9], the following holds:

$$P(h_{n+1}^{(j)} \notin \hat{H}_j) = \frac{\varepsilon}{\ell}.$$

We repeat this for all the observations in x_{n+1}. We define $\hat{H}$ as the set of all the sequences of length ℓ that can be generated by using elements from $\hat{H}_1$ as a first element, elements from $\hat{H}_2$ as a second one, and so on. Then we can derive the probability of error of our method as the probability of the correct sequence h_{n+1} of not being in the prediction set as:

$$P(h_{n+1} \notin \hat{H}) = P(h_{n+1}^{(1)} \notin \hat{H}_1 \vee h_{n+1}^{(2)} \notin \hat{H}_2 \vee ... \vee h_{n+1}^{(\ell)} \notin \hat{H}_\ell)$$

$$\leq \sum_{j=1}^{\ell} P(h_{n+1}^{(j)} \notin \hat{H}_j) = \ell\frac{\varepsilon}{\ell} = \varepsilon \qquad \blacksquare$$

Follows that $1 - \varepsilon$ is a lower–bound to the probability of error of the method.

B Datasets

B.1 *HMM-NORM* Dataset

We sampled 2000 sequences of length $\ell = 10$. The sequences were generated by using a continuous HMM with 3 hidden states, $S = \{s_1, s_2, s_3\}$, start probabilities $\Pi = \{0.6, 0.3, 0.1\}$, transition probabilities:

$$A = \{\alpha_{ij}\} = \begin{pmatrix} 0.7\ 0.2\ 0.1 \\ 0.3\ 0.5\ 0.2 \\ 0.3\ 0.3\ 0.4 \end{pmatrix},$$

and emission probabilities: $b_o s_1 \sim \mathcal{N}(-2, 0.7)$, $b_o s_2 \sim \mathcal{N}(0, 0.7)$, $b_o s_3 \sim \mathcal{N}(2, 0.7)$. Figure 6(a) graphically shows the distribution of $b_o s_1$, $b_o s_2$, and $b_o s_3$.

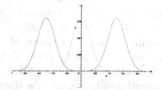
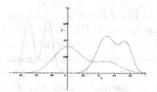

(a) Emission distributions for *HMM-NORM*.

(b) Emission distributions for *HMM-GMM*.

Fig. 6. Distribution of the emission probabilities for the three hidden states in *HMM-NORM* (left–hand figure), and in *HMM-GMM* (right–hand figure).

B.2 *HMM-GMM* Dataset

We sampled 2000 sequences of length $\ell = 10$. The sequences were generated by using a continuous HMM with 3 hidden states, $S = \{s_1, s_2, s_3\}$, start probabilities $\Pi = \{0.6, 0.3, 0.1\}$, transition probabilities:

$$A = \{\alpha_{ij}\} = \begin{pmatrix} 0.7\ 0.2\ 0.1 \\ 0.3\ 0.5\ 0.2 \\ 0.3\ 0.3\ 0.4 \end{pmatrix}.$$

Emission probabilities where given by one mixture of two Normal distributions. Let $\mathcal{G}(\mu, \sigma, w)$ be a mixture of two Normal distribution with means $\mu = (\mu_1, \mu_2)$, standard deviations $\sigma = (\sigma_1, \sigma_2)$, and weights $w = (w_1, w_2)$. That is:

$$\mathcal{G}(\mu, \sigma, w) = \sum_{i=1}^{2} w_i \mathcal{N}(\mu_i, \sigma_i).$$

The model we used had emission probabilities: $b_o s_1 \sim \mathcal{G}((0, 2), (0.7, 0.7), (0.7, 0.3))$, $b_o s_2 \sim \mathcal{G}((-2, -1), (0.25, 0.25), (0.5, 0.5))$, $b_o s_3 \sim \mathcal{G}((2, 3), (0.5, 0.3), (0.7, 0.3))$. Figure 6(b) graphically shows the distribution of $b_o s_1$, $b_o s_2$, and $b_o s_3$.

References

1. Forney Jr., G.D.: The Viterbi algorithm. Proc. IEEE **61**(3), 268–278 (1973)
2. Gammerman, A., Vovk, V.: Hedging predictions in machine learning. Comput. J. **50**(2), 151–163 (2007)
3. Melluish, T., Saunders, C., Nouretdinov, I., Vovk, V.: The typicalness framework: a comparison with the Bayesian approach. University of London, Royal Holloway (2001)
4. Rabiner, L.R.: A tutorial on hidden Markov models and selected applications in speech recognition. Proc. IEEE **77**(2), 257–286 (1989)
5. Seshadri, N., Sundberg, C.W.: List Viterbi decoding algorithms with applications. IEEE Trans. Commun. **42**(234), 313–323 (1994)
6. Shafer, G., Vovk, V.: A tutorial on conformal prediction. J. Mach. Learn. Res. **9**, 371–421 (2008)

7. Viterbi, A.J.: Error bounds for convolutional codes and an asymptotically optimum decoding algorithm. IEEE Trans. Inf. Theory **13**(2), 260–269 (1967)
8. Vovk, V., Fedorova, V., Nouretdinov, I., Gammerman, A.: Criteria of efficiency for conformal prediction. In: Gammerman, A., Luo, Z., Vega, J., Vovk, V. (eds.) COPA 2016. LNCS(LNAI), vol. 9653, pp. 23–39. Springer, Heidelberg (2016)
9. Vovk, V., Gammerman, A., Shafer, G.: Algorithmic Learning in a Random World. Springer, New York (2005)

Machine Learning

Variable Fidelity Regression Using Low Fidelity Function Blackbox and Sparsification

A. Zaytsev[1,2]([✉])

[1] IITP RAS, 127051 Moscow, Russia
likzet@gmail.com
[2] MIPT, Dolgoprudny, 141700 Moscow, Russia

Abstract. We consider construction of surrogate models based on variable fidelity samples generated by a high fidelity function (an exact representation of some physical phenomenon) and by a low fidelity function (a coarse approximation of the exact representation). A surrogate model is constructed to replace the computationally expensive high fidelity function. For such tasks Gaussian processes are generally used. However, if the sample size reaches a few thousands points, a direct application of Gaussian process regression becomes impractical due to high computational costs. We propose two approaches to circumvent this difficulty. The first approach uses approximation of sample covariance matrices based on the Nyström method. The second approach relies on the fact that engineers often can evaluate a low fidelity function on the fly at any point using some blackbox; thus each time calculating prediction of a high fidelity function at some point, we can update the surrogate model with the low fidelity function value at this point. So, we avoid issues related to the inversion of large covariance matrices — as we can construct model using only a moderate low fidelity sample size. We applied developed methods to a real problem, dealing with an optimization of the shape of a rotating disk.

Keywords: Multifidelity data · Gaussian process · Nonlinear regression · Nyström approximation · Cokriging

1 Introduction

Nowadays most advanced engineers encounter the problem of a surrogate model construction, when it is required to replace an expensive high fidelity function with an inexpensive but precise surrogate model [17]. Typically, to accomplish such a task one generates a sample of points and values of the corresponding high fidelity function at these points, and then using the generated sample and the machinery of regression analysis one constructs a surrogate model. Among various surrogate model construction techniques, the Gaussian process regression remains an attractive approach, as the machinery of this method provides a nonlinear regression model with prediction uncertainty estimate [17,37]). Moreover, Gaussian process framework provides straightforward solutions for classification

© Springer International Publishing Switzerland 2016
A. Gammerman et al. (Eds.): COPA 2016, LNAI 9653, pp. 147–164, 2016.
DOI: 10.1007/978-3-319-33395-3_11

[43], adaptive design of experiments [9] and surrogate based optimization [21] problems.

Another nice property of Gaussian process regression is the ability to treat variable fidelity data (see for example [12,16,22,27,28,35]): one can construct a surrogate model of a high fidelity function using data from high and low fidelity sources (e.g., a high fidelity function can be modeled by an experiment in a wind tunnel, and the low fidelity function can be realized by a computer simulation of the same physical process) and then use this model for surrogate-based optimization. Similar approaches are used for multiple output Gaussian processes modeling [2,8,11,25].

Straightforward maximum likelihood estimation of Gaussian process regression model parameters and application of model to new points require inversion of the covariance matrix of the sample [18]. The covariance matrix of the sample is a square matrix with number of both rows and columns equal to the sample size n. Consequently, as typically the covariance matrix has no specific structure, we need $O(n^2)$ to store the covariance matrix and $O(n^3)$ to invert it. Due to this computational complexity usually not more than a few thousands of points are used when training Gaussian Process regression. As a sample generated using the low fidelity function is often large, because the evaluation of a low fidelity function is significantly cheaper than that of a high fidelity function, the problem is even worse for variable fidelity data.

Currently there are several ways to avoid inversion of the full covariance matrix in Gaussian process regression. Using of Nyström approximation [13] of the covariance matrix has remained a popular approach to do large sample Gaussian process regression inference for more than 10 years [18,36,41]. The idea is to select a subsample of the full sample for which we can do Gaussian process regression inference, and then approximate the full sample covariance matrix and inverse of the full sample covariance matrix by combination of the covariance matrix for selected subsample and covariance between points in the selected subsample and in the full sample. Another approach consists of usage of Bayesian approximate inference to estimate the full sample likelihood with an easy-to-calculate expression [24,42]. Rather popular approach with proved theoretical properties is covariance tapering [19,38]: we suppose that covariance function equals zero for points with distance above the taper parameters, so we obtain sparse covariance matrices, and can proceed them with routines specific to sparse matrices. Hierarchical models move away the computational burden, as they split the sample to separate subsamples, which leads to covariance matrix with specific structure [5,33,39]. However, exact inference is possible if data have some specific structure: for example, [6] has developed an exact inference scheme to construct Gaussian process regression. Another example that works with variable fidelity data of big size with specific structure (we aggregate many low fidelity uncalibrated models using observations at the same points) was presented in [11]. However, as far as we know there are no approaches to large scale variable fidelity Gaussian process regression for data without any specific structure.

Another issue with Gaussian process regression lies in its bad extrapolation properties, since the model prediction at a new point is the weighted sum of values at given training points with weights defined by covariances between points [37]; i.e., the prediction can be determined only locally near the training points, and we need to be careful with points that are far away from the training sample.

We propose two approaches that mitigate the sample size limitation and improve the extrapolation properties of variable fidelity Gaussian process regression. The first approach uses the Nyström approximation to the covariance matrices and relies on the results obtained for a single fidelity data in the Sparse Gaussian process regression framework [18]. The main idea of the second approach is to use the low fidelity function blackbox during the model evaluation, so one can evaluate a low fidelity function on the fly only at the points where it is required to approximate a high fidelity function and use these evaluations to update the surrogate model predictions. While, for simple heuristic models it is a common practice to use a low fidelity function blackbox [1,29,40,44,45], Gaussian process regression doesn't support usage of such an approach in a direct way. As we are able to evaluate the low fidelity function at any point from the design space, we avoid usage of large sample to cover all the design space. Instead, it is sufficient only to get enough points to estimate parameters of Gaussian process regression model.

For proposed approaches we investigate their computational complexity and compare their accuracy using real and artificial data. The real problem at hand is optimization of a rotating disk in an aircraft engine. The problem of a disk shape optimization remains challenging and often involves usage of surrogate modeling [15,26], so it is required to construct accurate surrogate models for maximal stress and radial displacement of the disk used then for surrogate optimization. We compare four approaches to construct the rotating disk surrogate models: Gaussian process (kriging), Gaussian process for variable fidelity data (cokriging) and our approaches — Gaussian process for variable fidelity data with usage of a low fidelity blackbox and large scale variable fidelity Gaussian process regression.

The paper is organized as follows:

- Section 2 describes the Gaussian process regression framework;
- Section 3 outlines the Variable fidelity Gaussian process regression framework;
- Section 4 proposes an approach to construct Sparse Gaussian process regression for variable fidelity data;
- Section 5 describes our approach to Variable Fidelity Gaussian process regression with a low fidelity function blackbox;
- Section 6 provides the results of computational experiments for both real and artificial data,
- Conclusions are given in Sect. 7.

In Appendix we provide proofs of some technical statements and details on low and high fidelity models for rotating disk problem.

2   Gaussian Process Regression for a Single Fidelity Data

We consider a single training sample $D = (\mathbf{X}, \mathbf{y}) = \{\mathbf{x}_i, y_i = y(\mathbf{x}_i)\}_{i=1}^n$, where points $\mathbf{x} \in \mathbb{X} \subseteq \mathbb{R}^d$ and a function value $y(\mathbf{x}) \in \mathbb{R}$. We assume that $y(\mathbf{x}) = f(\mathbf{x}) + \varepsilon$, where $f(\mathbf{x})$ is a realization of a Gaussian process, and ε is a Gaussian white noise with variance σ^2. The goal is to construct a surrogate model for the target function $f(\mathbf{x})$.

The mean value and the covariance function

$$k(\mathbf{x}, \mathbf{x}') = \mathrm{cov}(f(\mathbf{x}), f(\mathbf{x}')) = \mathbb{E}\left(f(\mathbf{x}) - \mathbb{E}(f(\mathbf{x}))\right)\left(f(\mathbf{x}') - \mathbb{E}(f(\mathbf{x}'))\right)$$

completely define the Gaussian process $f(\mathbf{x})$. Without loss of generality we assume its mean value to be zero. We also assume that the covariance function belongs to some parametric family $\{k_{\boldsymbol{\theta}}(\mathbf{x}, \mathbf{x}'), \boldsymbol{\theta} \in \Theta \subseteq \mathbb{R}^p\}$; i.e., $k(\mathbf{x}, \mathbf{x}') = k_{\boldsymbol{\theta}}(\mathbf{x}, \mathbf{x}')$ for some $\boldsymbol{\theta} \in \Theta$. Thus $y(\mathbf{x})$ is also a Gaussian process [37] with zero mean and covariance function $\mathrm{cov}(y(\mathbf{x}), y(\mathbf{x}')) = k_{\boldsymbol{\theta}}(\mathbf{x}, \mathbf{x}') + \sigma^2 \delta(\mathbf{x} - \mathbf{x}')$, where $\delta(\mathbf{x} - \mathbf{x}')$ is the delta function. Example of a covariance function, widely used in applications, is the multivariate squared exponential covariance function [37] $k_{\boldsymbol{\theta}}(\mathbf{x}, \mathbf{x}') = \theta_0^2 \exp\left(-\sum_{k=1}^d \theta_k^2 (x_k - x_k')^2\right)$.

The covariance function parameters $\boldsymbol{\theta}$ and the variance σ^2 are known as fully specifying the data model. We use the Maximum Likelihood Estimation (MLE) of $\boldsymbol{\theta}$ and σ^2 [7,37] to fit the model; i.e., we maximize the logarithm of the training sample likelihood

$$\log p(\mathbf{y}|\mathbf{X}, \boldsymbol{\theta}, \sigma^2) = -\frac{1}{2}\left(n \log 2\pi + \log |\mathbf{K}| + \mathbf{y}^T \mathbf{K}^{-1} \mathbf{y}\right) \to \max_{\boldsymbol{\theta}, \sigma^2}, \tag{1}$$

where $\mathbf{K} = \{k_{\boldsymbol{\theta}}(\mathbf{x}_i, \mathbf{x}_j) + \sigma^2 \delta(\mathbf{x}_i - \mathbf{x}_j)\}_{i,j=1}^n$ is the matrix of covariances between values $\mathbf{y}(\mathbf{X})$ of the training sample and $|\mathbf{K}|$ is the determinant of $\mathbf{K}$. σ^2 plays the role of a regularization parameter for the kernel matrix $\{k_{\boldsymbol{\theta}}(\mathbf{x}_i, \mathbf{x}_j)\}_{i,j=1}^n$, being a matrix of covariances between values $f(\mathbf{X})$. The recent theoretical work [10] and the experimental works [4,46] suggest that, under general assumptions, MLE parameters estimates $\hat{\boldsymbol{\theta}}$ are accurate even if the sample size is limited and the model is misspecified.

Using estimates of $\boldsymbol{\theta}$ and σ^2 we can calculate the posterior mean and the covariances of $y(\mathbf{x})$ at new points playing, respectively, the role of a prediction and its uncertainty. The posterior mean $\mathbb{E}(\mathbf{y}(\mathbf{X}^*)|\mathbf{y}(\mathbf{X}))$ at the new points $\mathbf{X}^* = \{\mathbf{x}_i^*\}_{i=1}^{n^*}$ has the form

$$\hat{\mathbf{y}}(\mathbf{X}^*) = \mathbf{K}(\mathbf{X}^*, \mathbf{X})\mathbf{K}^{-1}\mathbf{y}, \tag{2}$$

where $\mathbf{K}(\mathbf{X}^*, \mathbf{X}) = \{k(\mathbf{x}_i^*, \mathbf{x}_j)\}_{i=1,\dots,n^*, j=1,\dots,n}$ are the covariances between the values $\mathbf{y}(\mathbf{X}^*)$ and $\mathbf{y}(\mathbf{X})$. The posterior covariance matrix $\mathbb{V}(\mathbf{X}^*) = \mathbb{E}\big[(\mathbf{y}(\mathbf{X}^*) - \mathbb{E}\mathbf{y}(\mathbf{X}^*))^T(\mathbf{y}(\mathbf{X}^*) - \mathbb{E}\mathbf{y}(\mathbf{X}^*)) \mid \mathbf{y}(\mathbf{X})\big]$ has the form

$$\mathbb{V}(\mathbf{X}^*) = \mathbf{K}(\mathbf{X}^*, \mathbf{X}^*) - \mathbf{K}(\mathbf{X}^*, \mathbf{X})\mathbf{K}^{-1}\mathbf{K}(\mathbf{X}, \mathbf{X}^*), \tag{3}$$

where $\mathbf{K}(\mathbf{X}^*, \mathbf{X}^*) = \{k(\mathbf{x}_i^*, \mathbf{x}_j^*) + \sigma^2 \delta(\mathbf{x}_i^* - \mathbf{x}_j^*)\}_{i,j=1}^{n^*}$ is the matrix of covariances between values $\mathbf{y}(\mathbf{X}^*)$.

3   Variable Fidelity Gaussian Process Regression

Now we consider the case of variable fidelity data: there are a sample of the low fidelity function $D_l = (\mathbf{X}_l, \mathbf{y}_l) = \left\{\mathbf{x}_i^l, y_l(\mathbf{x}_i^l)\right\}_{i=1}^{n_l}$ and a sample of the high fidelity function $D_h = (\mathbf{X}_h, \mathbf{y}_h) = \left\{\mathbf{x}_i^h, y_h(\mathbf{x}_i^h)\right\}_{i=1}^{n_h}$ with $\mathbf{x}_i^l, \mathbf{x}_i^h \in \mathbb{R}^d$, $y_l(\mathbf{x}), y_h(\mathbf{x}) \in \mathbb{R}$. The low fidelity function $y_l(\mathbf{x})$ and the high fidelity function $y_h(\mathbf{x})$ model the same physical phenomenon, but with different fidelities.

With the use of samples of the low and the high fidelity functions our aim is to construct, as accurately as possible, a surrogate model $\hat{y}_h(\mathbf{x}) \approx y_h(\mathbf{x})$ of the high fidelity function; moreover, we also need an uncertainty estimate of the prediction.

If data come from two sources of different fidelities, then an appropriate model should be used. We assume that the following variable fidelity data model holds true [16]:

$$y_l(\mathbf{x}) = f_l(\mathbf{x}) + \varepsilon_l, \ y_h(\mathbf{x}) = \rho y_l(\mathbf{x}) + y_d(\mathbf{x}),$$

where $y_d(\mathbf{x}) = f_d(\mathbf{x}) + \varepsilon_d$. $f_l(\mathbf{x})$, $f_d(\mathbf{x})$ are realizations of independent Gaussian processes with zero means and covariance functions $k_l(\mathbf{x}, \mathbf{x}')$ and $k_d(\mathbf{x}, \mathbf{x}')$, respectively, and ε_l, ε_d are Gaussian white noise processes with variances σ_l^2 and σ_d^2, respectively. We also set $\mathbf{X} = \begin{pmatrix} \mathbf{X}_l \\ \mathbf{X}_h \end{pmatrix}$, $\mathbf{y} = \begin{pmatrix} \mathbf{y}_l \\ \mathbf{y}_h \end{pmatrix}$. Then the posterior mean of the high-fidelity values at new points has the form

$$\hat{\mathbf{y}}_h(\mathbf{X}^*) = \mathbf{K}(\mathbf{X}^*, \mathbf{X})\mathbf{K}^{-1}\mathbf{y}, \tag{4}$$

where

$$\mathbf{K}(\mathbf{X}^*, \mathbf{X}) = \left(\rho\mathbf{K}_l(\mathbf{X}^*, \mathbf{X}_l) \ \rho^2\mathbf{K}_l(\mathbf{X}^*, \mathbf{X}_h) + \mathbf{K}_d(\mathbf{X}^*, \mathbf{X}_h)\right),$$

$$\mathbf{K}(\mathbf{X}, \mathbf{X}) = \begin{pmatrix} \mathbf{K}_l(\mathbf{X}_l, \mathbf{X}_l) & \rho\mathbf{K}_l(\mathbf{X}_l, \mathbf{X}_h) \\ \rho\mathbf{K}_l(\mathbf{X}_h, \mathbf{X}_l) & \rho^2\mathbf{K}_l(\mathbf{X}_h, \mathbf{X}_h) + \mathbf{K}_d(\mathbf{X}_h, \mathbf{X}_h) \end{pmatrix},$$

$\mathbf{K}_l(\mathbf{X}_a, \mathbf{X}_b)$, $\mathbf{K}_d(\mathbf{X}_a, \mathbf{X}_b)$ are the matrices of pairwise covariances for the Gaussian processes $y_l(\mathbf{x})$ and $y_d(\mathbf{x})$ for points from some samples $\mathbf{X}_a$ and $\mathbf{X}_b$, respectively. The posterior covariance matrix is as follows:

$$\mathbb{V}(\mathbf{X}^*) = \rho^2\mathbf{K}_l(\mathbf{X}^*, \mathbf{X}^*) + \mathbf{K}_d(\mathbf{X}^*, \mathbf{X}^*) - \mathbf{K}(\mathbf{X}^*, \mathbf{X})\mathbf{K}^{-1}\left(\mathbf{K}(\mathbf{X}^*, \mathbf{X})\right)^T. \tag{5}$$

To estimate covariance function parameters and noise variances for Gaussian processes $f_l(\mathbf{x})$ and $f_d(\mathbf{x})$ we use the following common algorithm [16]:

1. Estimate the parameters of the covariance function $k_l(\mathbf{x}, \mathbf{x})$ using the algorithm from Sect. 2 with sample $D = D_l$,
2. Calculate the posterior mean estimates $\hat{y}_l(\mathbf{x})$ of the Gaussian process $y_l(\mathbf{x})$ for $\mathbf{x} \in \mathbf{X}_h$,
3. Estimate the parameters of the Gaussian process $y_d(\mathbf{x})$ with the covariance function $k_d(\mathbf{x}, \mathbf{x}')$ and parameter ρ by maximizing likelihood (1) with $D = D_{\text{diff}} = (\mathbf{X}_h, \mathbf{y}_d = \mathbf{y}_h - \rho\hat{\mathbf{y}}_l(\mathbf{X}_h))$ and $k(\mathbf{x}, \mathbf{x}') = k_d(\mathbf{x}, \mathbf{x}')$.

As we have big enough sample of low fidelity data, we assume that we can get precise estimates of parameters of covariance function $k_l(\mathbf{x}, \mathbf{x})$, so we don't need to refine these estimates using high fidelity data.

4   Sparse Gaussian Process Regression

To perform inference for Variable Fidelity Gaussian process regression we have to invert the sample covariance matrix of size $n \times n$, where $n = n_h + n_l$. This operation is of complexity $O(n^3)$, so for samples of sizes larger than few thousands points we cannot construct a Gaussian process regression in a reasonable time.

In order to construct a Gaussian process regression for large sample sizes we propose to use an approximation to the exact inference. The Nyström approximation [18] of all involved matrices $\mathbf{K}(\mathbf{X}^*, \mathbf{X})$, $\mathbf{K}$ and $\mathbf{K}(\mathbf{X}^*, \mathbf{X}^*)$ allows one to obtain such an approximation.

Let us select from the initial sample a subsample $\mathbf{X}^1 = \begin{pmatrix} \mathbf{X}_l^1 \\ \mathbf{X}_h^1 \end{pmatrix}, \mathbf{y}^1 = \begin{pmatrix} \mathbf{y}_l(\mathbf{X}_l^1) \\ \mathbf{y}_h(\mathbf{X}_h^1) \end{pmatrix}$ of *base* points with the size $n_1 = n_h^1 + n_l^1$ to be small enough so we can perform an exact inference for it. The simplest, rather robust and efficient way for this is to perform uniform random selection without repetitions among points from the initial samples.

Hence, by definition,

$$\mathbf{K}_{11} = \begin{pmatrix} \mathbf{K}_l(\mathbf{X}_l^1, \mathbf{X}_l^1) & \rho \mathbf{K}_l(\mathbf{X}_l^1, \mathbf{X}_h^1) \\ \rho \mathbf{K}_l(\mathbf{X}_h^1, \mathbf{X}_l^1) & \rho^2 \mathbf{K}_l(\mathbf{X}_h^1, \mathbf{X}_h^1) + \mathbf{K}_d(\mathbf{X}_h^1, \mathbf{X}_h^1) \end{pmatrix},$$

$$\mathbf{K}_1 = \begin{pmatrix} \mathbf{K}_l(\mathbf{X}_l^1, \mathbf{X}_l) & \rho \mathbf{K}_l(\mathbf{X}_l^1, \mathbf{X}_h) \\ \rho \mathbf{K}_l(\mathbf{X}_h^1, \mathbf{X}_l) & \rho^2 \mathbf{K}_l(\mathbf{X}_h^1, \mathbf{X}_h) + \mathbf{K}_d(\mathbf{X}_h^1, \mathbf{X}_h) \end{pmatrix},$$

$$\mathbf{K}_1^* = \begin{pmatrix} \rho \mathbf{K}_l(\mathbf{X}^*, \mathbf{X}_l^1) & \rho^2 \mathbf{K}_l(\mathbf{X}^*, \mathbf{X}_h^1) + \mathbf{K}_d(\mathbf{X}^*, \mathbf{X}_h^1) \end{pmatrix}$$

for some new points $\mathbf{X}^* = \{\mathbf{x}_i^*\}_{i=1}^{n^*}$ and so using the Nyström approximation we get approximations of the matrices $\mathbf{K}(\mathbf{X}^*, \mathbf{X})$, $\mathbf{K}$ and $\mathbf{K}(\mathbf{X}^*, \mathbf{X}^*)$, respectively:

$$\hat{\mathbf{K}}(\mathbf{X}^*, \mathbf{X}) = \mathbf{K}_1^* \mathbf{K}_{11}^{-1} \mathbf{K}_1, \quad \hat{\mathbf{K}} = (\mathbf{K}_1)^T \mathbf{K}_{11}^{-1} \mathbf{K}_1, \quad \hat{\mathbf{K}}(\mathbf{X}^*, \mathbf{X}^*) = \mathbf{K}_1^* \mathbf{K}_{11}^{-1} (\mathbf{K}_1^*)^T.$$

We set

$$\mathbf{R} = \begin{pmatrix} \frac{1}{\sigma_l} \mathbf{I}_{n_l} & 0 \\ 0 & \frac{1}{\sqrt{\rho^2 \sigma_l^2 + \sigma_d^2}} \mathbf{I}_{n_h} \end{pmatrix},$$

where $\mathbf{I}_k$ is the identity matrix of size k, $\mathbf{C}_1 = \mathbf{R}\mathbf{K}_1$ and $\mathbf{V} = \mathbf{C}_1 \mathbf{V}_{11}^{-T}$, $\mathbf{V}_{11}$ is the Cholesky decomposition of $\mathbf{K}_{11}$.

Theorem 1. *For the posterior mean and the posterior covariance matrix the following Nystrom approximations hold*

$$\hat{\mathbf{y}}_h^{Ny}(\mathbf{X}^*) = \mathbf{K}_1^* \mathbf{V}_{11}^{-1} (\mathbf{I}_{n_1} + \mathbf{V}^T \mathbf{V})^{-1} \mathbf{V}^T \mathbf{R} \mathbf{y}, \tag{6}$$

$$\mathbb{V}^{Ny}(\mathbf{X}^*) = \mathbf{K}_1^* \mathbf{V}_{11}^{-1} (\mathbf{I}_{n_1} + \mathbf{V}^T \mathbf{V})^{-1} \mathbf{V}_{11}^{-T} \mathbf{K}_1^{*T} + (\rho^2 \sigma_l^2 + \sigma_d^2) \mathbf{I}_{n^*}. \tag{7}$$

Theorem 2. *The computational complexities of the posterior mean and the posterior covariance matrix calculation using (6) and (7) at one point are $O(nn_1^2)$.*

Proof of these theorems are in Appendix A.

5   Gaussian Process Regression for Multifidelity Data with Blackbox for Low Fidelity Function

Suppose that we have a blackbox for the low fidelity function $y_l(\mathbf{x})$; i.e., the blackbox estimates the low fidelity function value at any point from the design space $\mathbb{X} \subseteq \mathbb{R}^d$ on the fly. Let us assume that we have already constructed a Variable fidelity Gaussian processes surrogate model and can calculate predictions using (4) and (5). We can't use huge sample of low fidelity function values at corresponding points due to typical computational limitations for Gaussian process regression. Instead, in order to improve an accuracy of these predictions we can update the posterior mean and the posterior variance of $y_h(\mathbf{x})$ at a new point $\mathbf{x}$ with the low fidelity function value $y_l(\mathbf{x})$ at this point, as calculated by the blackbox. Let us describe a computationally efficient procedure to calculate the update.

We set

$$\mathbf{k}_l(\mathbf{x}, \mathbf{X}) = \begin{pmatrix} \mathbf{K}_l(\mathbf{x}, \mathbf{X}_l) \\ \rho \mathbf{K}_l(\mathbf{x}, \mathbf{X}_h) \end{pmatrix},$$

where $\mathbf{x}$ is a some new point. For a sample with an additional point $\mathbf{x}$ included we get an expanded covariance matrix:

$$\mathbf{K}_{\text{exp}} = \begin{pmatrix} \mathbf{K} & \mathbf{k}_l \\ \mathbf{k}_l^T & k_l(\mathbf{x}, \mathbf{x}) \end{pmatrix}.$$

Suppose we know the Cholesky decompositions $\mathbf{L}$ and $\mathbf{L}^{-1}$ of the initial training sample covariance matrix $\mathbf{K}$ and its inverse $\mathbf{K}^{-1}$, respectively. To calculate the posterior mean and the posterior variance for the expanded model we will update these Cholesky decompositions and then update the posterior mean and the posterior variance values.

If we have an $n \times n$ matrix $\mathbf{K}_n$ and the Cholesky decomposition of it, we can get the updated Cholesky decomposition of the matrix $\mathbf{K}_{n+1}$ of size $(n+1) \times (n+1)$ if the initial matrix is in the upper left corner of the new matrix $\mathbf{K}_{n+1}$ with computational complexity $O(n^2)$ using a common routine [20]. To update inverse of the Cholesky decomposition we also need $O(n^2)$ operations, as it differs from the initial Cholesky decomposition only in the last row and is lower triangular. Therefore, we can calculate the matrix $\mathbf{K}_{\text{exp}}^{-1}$ in $O(n^2)$ operations.

The expanded vector of covariances between the new point $\mathbf{x}$ and the initial training sample has the form

$$\mathbf{k}_{\text{exp}} = \begin{pmatrix} \rho \mathbf{K}_l(\mathbf{x}, \mathbf{X}_l) \\ \rho^2 \mathbf{K}_l(\mathbf{x}, \mathbf{X}_h) + \mathbf{K}_d(\mathbf{x}, \mathbf{X}_h) \\ \rho k_l(\mathbf{x}, \mathbf{x}) \end{pmatrix}.$$

Using the value $y_l(\mathbf{x})$ calculated by the blackbox, we set $\mathbf{y}_{\text{exp}} = \left(\mathbf{y}^T, y_l(\mathbf{x})\right)^T$. Then the updated expressions for the posterior mean and the posterior variance are as follows:

$$\hat{y}_h^{\text{exp}}(\mathbf{x}) = \mathbf{k}_{\text{exp}} \mathbf{K}_{\text{exp}}^{-1} \mathbf{y}_{\text{exp}}, \tag{8}$$

$$\mathbb{V}_{\text{exp}}(\mathbf{x}) = \rho^2 \mathbf{K}_l(\mathbf{x}, \mathbf{x}) + \mathbf{K}_d(\mathbf{x}, \mathbf{x}) - \mathbf{k}_{\text{exp}}^T \mathbf{K}_{\text{exp}}^{-1} \mathbf{k}_{\text{exp}}. \tag{9}$$

As the Cholesky decomposition for the updated model differs only in the last row we can calculate (8) and (9) in $O(n^2)$ operations.

The total computational complexity is the sum of the computational complexities of the Cholesky decomposition update and the posterior mean and the posterior variance recalculation, so for a Variable fidelity Gaussian process regression with a blackbox, representing the low fidelity function, the following assertions holds.

Theorem 3. *Suppose we know the Cholesky decompositions* $\mathbf{L}$ *and* $\mathbf{L}^{-1}$ *of the initial training sample covariance matrix* $\mathbf{K}$ *and its inverse* $\mathbf{K}^{-1}$, *respectively. Then we can calculate the posterior mean* $\hat{y}_h^{\exp}(\mathbf{x})$ *via* (8) *and the variance* $\mathbb{V}_{\exp}(\mathbf{x})$ *via* (9) *in* $O(n^2)$ *operations, where* $n = n_l + n_h$.

As we add only one point to the initial training sample, we expect that estimate of parameters of Gaussian processes model remains accurate enough. While it can be reasonable to add many points in some cases, this issue raises the complex question on how and when we should re-estimate Gaussian processes parameters as we add more points. Using blackbox for the low fidelity function we can get significantly more accurate approximation with small additional computational cost.

6   Numerical Examples

In this section we consider several problems: two artificial problems and a real applied problem of surrogate model construction for a rotating disk from an aircraft engine. We compare the four approaches below for a surrogate model construction; two latter approaches are introduced above:

- GP — Gaussian Process Regression using only high fidelity data,
- VFGP — Variable Fidelity Gaussian Process Regression using high and low fidelity data,
- SVFGP — Sparse VFGP, which is a version of VFGP for the case of large training samples introduced in Sect. 4,
- BB VFGP — VFGP with the low fidelity function realized by a black box introduced in Sect. 5. In experiments we use the same design of experiments as in case of VFGP, while for model update for each new point we use low fidelity function value at this point.

As a covariance function for a Gaussian process regression we use the multivariate squared exponential covariance function, see [37]. To regularize the problem and avoid inversion of large ill-conditioned matrices, we impose a prior distribution of nugget term in Bayesian way [7], so we are sure that for all four approaches we avoid problems with poor estimation of parameters for Gaussian Processes for large samples due to computational issues (linked with small values of regularization parameter σ^2 (nugget effect) [31,34]). To estimate parameters in SVFGP we use only a selected subsample of points, while we use the full sample to predict values at new points.

To measure the accuracy of the obtained surrogate models we use an RRMS error estimated by k-fold cross-validation procedure [23] if not specified otherwise. Note that we use low fidelity point for training only if the same point doesn't belong to the selected high fidelity test design if not specified otherwise. For a single target variable and a test sample $D_{\text{test}} = \{\mathbf{x}_i^{\text{test}}, y_i^{\text{test}} = f_h(\mathbf{x}_i^{\text{test}})\}_{i=1}^{n_t}$ the RRMS error for a surrogate model $\hat{y}(\mathbf{x})$ equals to

$$RRMS(D_{\text{test}}, \hat{y}) = \sqrt{\frac{\sum_{i=1}^{n_t}(\hat{y}_h(\mathbf{x}_i^{\text{test}}) - y_i^{\text{test}})^2}{\sum_{i=1}^{n_t}(\overline{y} - y_i^{\text{test}})^2}},$$

here $\overline{y} = \frac{1}{n_t}\sum_{i=1}^{n_t} y_i^{\text{test}}$. The value of the RRMS error typically lies between 0 and 1. Accurate models have RRMS values close to 0, while inaccurate models have RRMS values close to or greater than 1.

6.1 Artificial Problem with Big Sample Size

To benchmark proposed approaches we use an artificial function with multiple local peculiarities and input dimension $d = 6$, so we really need rather big sample to get an accurate surrogate model. As a high fidelity function $y_h(\mathbf{x})$ and a low fidelity function $y_l(\mathbf{x})$ we use

$$y_h(\mathbf{x}) = 20 + \sum_{i=1}^{d}(x_i^2 - 10\cos(2\pi x_i)) + \varepsilon_h, \ \mathbf{x} \in [0,1]^d,$$

$$y_l(\mathbf{x}) = y_h(\mathbf{x}) + 0.2\sum_{i=1}^{d}(x_i + 1)^2 + \varepsilon_l, \ \mathbf{x} \in [0,1]^d.$$

The high fidelity function was corrupted by a Gaussian white noise ε_h with variance 0.001, and the low fidelity function was corrupted by a Gaussian white noise ε_l with variance 0.002. When preparing samples for experiments we generate points in $[0,1]^d$ using Latin Hypercube Sampling [32]. To test extrapolation properties we limit training sample points to the region with range $[0, 0.5]$ instead of $[0, 1]$ for one of 6 input variables. The high fidelity sample size was $n_h = 100$ and the size of the subsample for SVFGP was $n_l^1 = 1000$ in all experiments.

The results were averaged over 5 runs for each considered value of n_l. We thus have

- Table 1 contains RRMS errors for VFGP, SVFGP, and BB VFGP,
- Table 2 contains RRMS errors for VFGP, SVFGP, and BB VFGP in case we use the surrogate model in extrapolation regime,
- Table 3 provides training times for VFGP, SVFGP and BB VFGP approach.

One can see that RRMS errors of SVFGP are comparable with RRMS errors of VFGP for the same sample size, while the training time of SVFGP is tremendously smaller when the sample size is equal to 5000, and for SVFGP the training time increases only slightly when the sample size increases. For BB VFGP training time in this experiment coincides with that of VFGP, while for 1000 training

Table 1. Comparison of RRMS errors

n_l	1000	3000	5000
VFGP	0.0502	0.0170	0.0058
SVFGP	0.0502	0.0305	0.0260
BB VFGP	0.0010	0.00029	0.00017

Table 2. Comparison of extrapolation RRMS errors

n_l	1000	3000	5000
VFGP	0.3636	0.1351	0.1028
SVFGP	0.3636	0.3281	0.3586
BB VFGP	0.000998	0.00113	0.00034

Table 3. Comparison of training times in seconds for Ubuntu PC, Intel-Core $i7$ with 4 physical cores, 3.4 GHz, 16 Gb RAM.

n_l	1000	3000	5000
VFGP	30.46	852.70	7283.27
SVFGP	30.46	33.42	37.50
BB VFGP	30.38	842.97	7672.60

points we get better results with BB VFGP than for 5000 training points and VFGP. If we calculate prediction in extrapolation regime, we get significantly better results with BB VFGP.

6.2   Rotating Disk Problem

Now let us compare the approaches to the construction of surrogate models on a real applied problem of rotating disk surrogate modeling.

Rotating Disk Model Description. A high speed rotating risk is an important part of an aircraft engine (see Fig. 1a), three parameters define quality of the disk: the mass of the disk, the maximal radial displacement u_{max}, the maximal stress s_{max} [3,6,30]. It is easy to calculate mass of the disk, as we know all geometrical parameters of the disk, while surrogate modeling of the maximal radial displacement and the maximal stress is challenging [26,30]. So the focus here is on modeling of the maximal radial displacement and the maximal stress.

Used parametrization of the rotating disk geometry consists of 8 parameters: the radii r_i, $i = 1, \ldots, 6$, which control where the thickness of the rotating disk changes, and the values t_1, t_3, t_5, which control the corresponding changes in thickness. In the considered surrogate modeling problem we fix the radii r_4, r_5 and the thickness t_3 of a rotating disk, so the input dimension for the surrogate

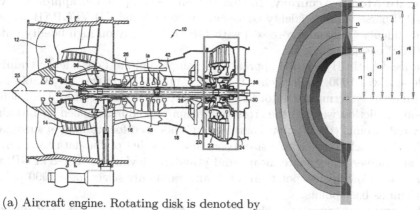

(a) Aircraft engine. Rotating disk is denoted by the bold rectangle at the right side of the figure.

(b) Rotating disk geometry

Fig. 1. Rotating disk problem

model is 6. The geometry and the parametrization of the rotating disk are shown in Fig. 1b.

There are two available solvers for u_{max} and s_{max} calculation. The low fidelity function is calculated using Ordinary Differential Equations (ODE) solver based on a simple Runge–Kutta's method. The high fidelity function is calculated using Finite Element Model (FEM) solver. A single evaluation of the low fidelity function takes ~ 0.01 s, and a single evaluation of the high fidelity function takes ~ 300 s. More detailed comparison of solvers is given in Appendix B.

Table 4. RRMS errors for introduced approaches with standard deviations

Output u_{max}				
n_h	20	40	60	80
GP	0.287 ± 0.039	0.143 ± 0.031	0.082 ± 0.020	0.095 ± 0.023
VFGP	0.212 ± 0.075	0.088 ± 0.009	0.064 ± 0.007	0.068 ± 0.006
SVFGP	0.125 ± 0.029	0.074 ± 0.016	0.041 ± 0.007	0.047 ± 0.011
BB VFGP	0.123 ± 0.019	0.053 ± 0.008	0.030 ± 0.007	0.034 ± 0.006
Output s_{max}				
n_h	20	40	60	80
GP	0.505 ± 0.10	0.367 ± 0.15	0.251 ± 0.049	0.196 ± 0.014
VFGP	0.363 ± 0.07	0.261 ± 0.06	0.193 ± 0.011	0.123 ± 0.043
SFGP	0.190 ± 0.06	0.122 ± 0.06	0.119 ± 0.015	0.088 ± 0.027
BB VFGP	0.158 ± 0.03	0.162 ± 0.03	0.137 ± 0.024	0.078 ± 0.020

Surrogate Model Accuracy. In this section we compare our approaches via SVFGP (Sparse variable fidelity Gaussian processes) and BB VFGP (Blackbox variable fidelity Gaussian processes) with GP (based only on high fidelity data) and VFGP baseline methods.

We used Latin Hypercube approach to sample points. Low fidelity training sample size was 1000, High fidelity training sample size n_h was 20, 40, 60, and 80 in different experiments. In order to estimate the accuracy of a high fidelity function prediction we used the cross-validation procedure, applied to 140 high fidelity data points (these points contain n_h points used for training of surrogate models). For each fixed sample size n_h we used 5 splits of the data to training and test samples to estimate means and standard deviations. For SVFGP, we use $n_l = 5000$ low fidelity points in total, and randomly select $n_l^1 = 1000$ points from them as base points.

The results are given in Table 4 for $u_{\max}$ and $s_{\max}$ outputs: VFGP outperforms GP, and both SVFGP and BB VFGP outperform VFGP in terms of RRMS error. Therefore, we decide which one to use, SVFGP or BB VFGP, by taking into account whether the blackbox for low fidelity function during a surrogate model usage is available, or whether one uses the surrogate model in extrapolation regime, etc.

6.3   Optimization of Rotating Disk Shape

We optimize the shape of the rotating disk described:

$$m, u_{\max} \rightarrow \min_{r_1,\dots,r_6,t_1,t_3,t_5}, \tag{10}$$

$$u_{\max} \leq 0.3, s_{max} \leq 600,$$
$$10 \leq r_1 \leq 110, 120 \leq r_2 \leq 140,$$
$$150 \leq r_3 \leq 168, 170 \leq r_4 \leq 200,$$
$$4 \leq t_1 \leq 50, 4 \leq t_3 \leq 50,$$
$$r_5 = 210, r_6 = 230, t_5 = 32.$$

The presented problem has multiple objectives, and we are looking for a Pareto frontier, not a single point.

Single optimization run is the following:

- Generate initial high fidelity sample D_h of 30 points using the Latin Hypercube sampling.
- Construct surrogate models using GP, VFGP, SVFGP and BB VFGP approaches using the generated high fidelity sample D_h and low fidelity sample D_l of size 1000 for GP, VFGP and BB VFGP and of size 5000 for SVFGP.
- Solve multiobjective optimization problem at hand using these surrogate models as the target functions and constraints.
- Calculate true values at Pareto frontiers obtained during optimization using high fidelity solver to estimate quality of models.

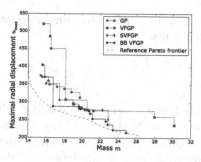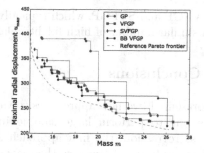

Fig. 2. Pareto frontiers obtained using optimization of surrogate models constructed with GP, VFGP, SVFGP and BB VFGP approaches along with the reference Pareto frontier

Table 5. Optimization results for different surrogate models along with minimal values for different optimization objectives. Also we present proportion of feasible points in the final Pareto frontier. The best values are in bold font.

Objective	GP	VFGP	SVFGP	BB VFGP
m	16.62	15.69	**15.09**	15.63
$0.8m + 0.2u_{\max}$	73.65	70.74	70.71	**68.10**
$0.6m + 0.4u_{\max}$	125.10	117.37	116.21	**112.55**
$0.4m + 0.6u_{\max}$	176.55	163.89	161.18	**156.99**
$0.2m + 0.8u_{\max}$	228.00	210.33	206.12	**201.44**
$u_{\max}$	279.44	256.77	251.05	**245.89**
Feasible points share	0.54	0.57	0.55	**0.75**

Due to properties of applied optimization algorithm sizes of Pareto frontiers can slightly differ for different runs of optimization algorithm, with mean size of Pareto frontier about 30 points [14]. So we need about 50 runs of high fidelity function to solve this optimization problem. In order to recover a reference Pareto frontier we constructed an accurate surrogate model using 5000 high fidelity points from uniform design over all the design space and additional sampling in a region where Pareto frontier points are located. So, instead of using a solver to evaluate an original function during optimization runs we used this surrogate model.

The examples of obtained Pareto frontiers for a single optimization run is in Fig. 2. For these runs SVFGP and BB VFGP work better than GP and VFGP.

Results of optimization are in Table 5. We compare minimum values of different weighted sums of two target variables m and $u_{\max}$ averaged over 10 runs of optimization for different initial samples. We obtain the best value of mass m output using SVFGP algorithm and the best value of $u_{\max}$ using BB VFGP algorithm while optimizations based on GP and VFGP work worse. Also, with BB VFGP we produce significantly larger amount of feasible points compared to

GP, VFGP and SVFGP, which typically leads to better Pareto frontier coverage with similar number of high fidelity blackbox runs.

7 Conclusions

We presented two new approaches to variable fidelity surrogate modeling, which allow one to perform large sample inference for Variable Fidelity Gaussian process regression: the first approach approximates the full covariance matrix of the sample and its inverse, the second approach uses the available low fidelity black box to update the surrogate model with the low fidelity function value at the point where one wants to estimate the high fidelity function thus avoiding requirement to use large low fidelity sample. Using developed approaches we can perform large sample inference for variable fidelity Gaussian process regression and construct more accurate surrogate models.

Acknowledgments. We thank Dmitry Khominich from DATADVANCE llc for making the solvers for rotating disk problem available, and Tatyana Alenkaya from MIPT for proofreading of the article. The research was conducted in IITP RAS and supported solely by the Russian Science Foundation grant (project 14-50-00150).

Appendix

A   Proof of Technical Statements

In this section we provide the proofs of the statements of Sect. 4.

Proof (Proof of Statement 1). For the posterior mean we get:

$$\hat{y}_h(x^*) \approx K_1^* K_{11}^{-1} K_1^T (K_1 K_{11}^{-1} K_1^T + R^{-2})^{-1} y = K_1^* K_{11}^{-1} K_1^T R (R K_1 K_{11}^{-1} K_1^T R + I_n)^{-1} R y =$$
$$= K_1^* K_{11}^{-1} C_1^T (C_1 K_{11}^{-1} C_1^T + I_n)^{-1} R y = K_1^* K_{11}^{-1} (C_1^T C_1 K_{11}^{-1} + I_{n_1})^{-1} C_1^T R y =$$
$$= K_1^* (C_1^T C_1 + K_{11})^{-1} C_1^T R y = K_1^* (C_1^T C_1 + V_{11}^T V_{11})^{-1} C_1^T R y =$$
$$= K_1^* V_{11}^{-1} (V_{11}^{-T} C_1^T C_1 V_{11}^{-1} + I_{n_1})^{-1} V_{11}^{-T} C_1^T R y = K_1^* V_{11}^{-1} (I_{n_1} + V^T V)^{-1} V^T R y.$$

We use the same approach to derive an equation for the posterior variance:

$$\mathbb{V}(X^*) - (\rho^2 \sigma_l^2 + \sigma_d^2) I_{n^*} \approx K_1^* K_{11}^{-1} K_1^{*T} - K_1^* K_{11}^{-1} K_1^T (R^{-2} + K_1 K_{11}^{-1} K_1^T)^{-1} K_1 K_{11}^{-1} K_1^{*T} =$$
$$= K_1^* (K_{11}^{-1} - K_{11}^{-1} K_1^T (R^{-2} + K_1 K_{11}^{-1} K_1^T)^{-1} K_1 K_{11}^{-1}) K_1^{*T} =$$
$$= K_1^* (K_{11} + K_1^T R^2 K_1)^{-1} K_1^{*T} = K_1^* (V_{11}^T V_{11} + C_1^T C_1)^{-1} K_1^{*T} =$$
$$= K_1^* V_{11}^{-1} (I_{n_1} + V^T V)^{-1} V_{11}^{-T} K_1^{*T}.$$

Proof (Proof of Statement 2). First of all we have to calculate the matrices V_{11} and $V = R K_1 V_{11}^{-T}$. The matrix V_{11} is of size $n_1 \times n_1$, so we need $O(n_1^3)$ to get its inverse. To calculate $K_1 V_{11}^{-T}$ we need $O(n_1^2 n)$ operations. Finally, as R is a diagonal matrix, we use $O(n_1 n)$ operations to get V.

In case $n^* = 1$ to get the posterior mean we have to calculate $\mathbf{V}_{11}(\mathbf{I}_{n_1} + \mathbf{V}^T\mathbf{V})^{-1}\mathbf{V}^T\mathbf{y}$. We use $O(n_1^2 n)$ operations to calculate $\mathbf{V}^T\mathbf{V}$, to inverse $\mathbf{I}_{n_1} + \mathbf{V}^T\mathbf{V}$ we need $O(n_1^3)$ operations, to calculate $\mathbf{V}_{11}(\mathbf{I}_{n_1} + \mathbf{V}^T\mathbf{V})^{-1}\mathbf{V}^T$ one uses extra $O(n_1^2 n)$ operations, and finally to calculate the posterior mean we need additional $O(n_1 n)$ operations. Consequently, to calculate the posterior mean we use $O(n_1^2 n)$ operations.

In the same way in order to calculate $\mathbf{V}_{11}(\mathbf{I}_{n_1} + \mathbf{V}^T\mathbf{V})^{-1}\mathbf{V}_{11}^{-1}$ we need $O(n_1^2 n)$ operations to calculate $(\mathbf{I}_{n_1} + \mathbf{V}^T\mathbf{V})^{-1}$ and additional $O(n_1^3)$ operations to get the final matrix. Consequently, in order to calculate the posterior variance we use $O(n_1^2 n)$ operations.

Finally, we need $O(n_1^2 n)$ operations to compute the required matrices, and $O(n_1^2 n)$, to obtain the posterior mean and the posterior variance from these precomputed matrices. So, the total computational complexity is $O(n_1^2 n)$.

B   Comparison of Low and High Fidelity Model for Rotating Disk

There are two available solvers for $u_{\max}$ and $s_{\max}$ calculation. The low fidelity function is calculated using Ordinary Differential Equations (ODE) solver based on a simple Runge–Kutta's method. The high fidelity function is calculated using Finite Element Model (FEM) solver from ANSYS.

To compare the solvers we draw the scatter plots of low and high fidelity values and also plot slices of the corresponding functions. We generate a random sample of points in a specified design space box, calculate the low and high fidelity function values and draw the low fidelity function values versus the high fidelity function values at the same points. The scatter plots are in Fig. 3: the difference between values increases significantly when the values are increasing.

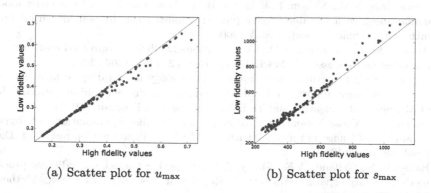

(a) Scatter plot for $u_{\max}$ (b) Scatter plot for $s_{\max}$

Fig. 3. Comparison of the high and the low fidelity solvers via scatter plots

For the central point of the design space box with $r_1 = 0.06, r_2 = 0.13, r_3 = 0.16, r_4 = 0.185, t_1 = 0.027, t_3 = 0.027$ we construct one-dimensional slices by

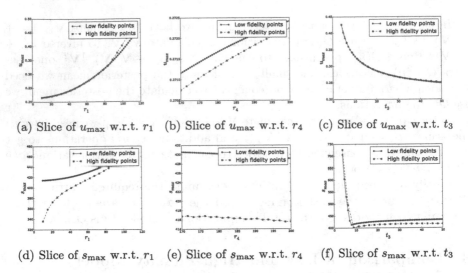

(a) Slice of $u_{\max}$ w.r.t. r_1 (b) Slice of $u_{\max}$ w.r.t. r_4 (c) Slice of $u_{\max}$ w.r.t. t_3

(d) Slice of $s_{\max}$ w.r.t. r_1 (e) Slice of $s_{\max}$ w.r.t. r_4 (f) Slice of $s_{\max}$ w.r.t. t_3

Fig. 4. Comparison of the high and the low fidelity solvers via outputs' slices

varying single input variable in specified bounds. Slices for different input variables for $u_{\max}$ and for $s_{\max}$ are given in Fig. 4. In case of $u_{\max}$ the high and the low fidelity functions have the same behaviour, and the low fidelity function models the high fidelity function accurately. For $s_{\max}$ the high and the low fidelity functions are sometimes different: their behaviours differ for a slice along r_1 input, and local maxima differ for slice along t_3 input.

References

1. Alexandrov, N.M., Nielsen, E.J., Lewis, R.M., Anderson, W.K.: First-order model management with variable-fidelity physics applied to multi-element airfoil optimization. Technical report, NASA (2000)
2. Álvarez, M.A., Lawrence, N.D.: Computationally efficient convolved multiple output Gaussian processes. J. Mach. Learn. Res. **12**, 1425–1466 (2011)
3. Armand, S.C.: Structural optimization methodology for rotating disks of aircraft engines. Technical report, National Aeronautics and Space Administration, Office of Management, Scientific and Technical Information Program (1995)
4. Bachoc, F.: Cross validation and maximum likelihood estimations of hyperparameters of Gaussian processes with model misspecification. Comput. Stat. Data Anal. **66**, 55–69 (2013)
5. Banerjee, S., Gelfand, A.E., Finley, A.O., Sang, H.: Gaussian predictive process models for large spatial data sets. J. Royal Stat. Soc. Ser. B (Statist. Method.) **70**(4), 825–848 (2008)
6. Belyaev, M., Burnaev, E., Kapushev, Y.: Gaussian process regression for structured data sets. In: Gammerman, A., Vovk, V., Papadopoulos, H. (eds.) SLDS 2015. LNCS, vol. 9047, pp. 106–115. Springer, Heidelberg (2015)
7. Bishop, C.M.: Pattern Recognition and Machine Learning. Springer, New York (2006)

8. Boyle, P., Frean, M.: Dependent Gaussian processes. Adv. Neural Inf. Process. Syst. **17**, 217–224 (2005)
9. Burnaev, E., Panov, M.: Adaptive design of experiments based on Gaussian processes. In: Gammerman, A., Vovk, V., Papadopoulos, H. (eds.) SLDS 2015. LNCS, vol. 9047, pp. 116–125. Springer, Heidelberg (2015)
10. Burnaev, E.V., Zaytsev, A.A., Spokoiny, V.G.: The Bernstein-von Mises theorem for regression based on Gaussian processes. Russ. Math. Surv. **68**(5), 954–956 (2013)
11. Chang, W., Haran, M., Olson, R., Keller, K., et al.: Fast dimension-reduced climate model calibration and the effect of data aggregation. Ann. Appl. Stat. **8**(2), 649–673 (2014)
12. Doyen, P.: Porosity from seismic data: a geostatistical approach. Geophysics **53**(10), 1263–1275 (1988)
13. Drineas, P., Mahoney, M.W.: On the Nyström method for approximating a Gram matrix for improved kernel-based learning. J. Mach. Learn. Res. **6**, 2153–2175 (2005)
14. Druot, T., Alestra, S., Brand, C., Morozov, S.: Multi-objective optimization of aircrafts family at conceptual design stage. In: Design and Optimization Symposium. Albi, France, In Inverse Problems (2013)
15. Farshi, B., Jahed, H., Mehrabian, A.: Optimum design of inhomogeneous non-uniform rotating discs. Comput. Struct. **82**(9), 773–779 (2004)
16. Forrester, A.I.J., Sóbester, A., Keane, A.J.: Multi-fidelity optimization via surrogate modelling. Proc. Roy. Soc. A Math. Phys. Eng. Sci. **463**(2088), 3251–3269 (2007)
17. Forrester, A.I.J., Sóbester, A., Keane, A.J.: Engineering Design Via Surrogate Modelling: a Practical Guide. J. Wiley, Chichester (2008)
18. Foster, L., Waagen, A., Aijaz, N., Hurley, M., Luis, A., Rinsky, J., Satyavolu, C., Way, M.J., Gazis, P., Srivastava, A.: Stable and efficient Gaussian process calculations. J. Mach. Learn. Res. **10**, 857–882 (2009)
19. Furrer, R., Genton, M.G., Nychka, D.: Covariance tapering for interpolation of large spatial datasets. J. Comput. Graphical Stat. **15**(3), 502–523 (2006)
20. Golub, G.H., Van Loan, C.F.: Matrix Computations, vol. 3. JHU Press, Baltimore (2012)
21. Grihon, S., Burnaev, E., Belyaev, M., Prikhodko, P.: Surrogate modeling of stability constraints for optimization of composite structures. In: Koziel, S., Leifsson, L. (eds.) Surrogate-Based Modeling and Optimization, pp. 359–391. Springer, New York (2013)
22. Han, Z., Görtz, S., Zimmermann, R.: Improving variable-fidelity surrogate modeling via gradient-enhanced kriging and a generalized hybrid bridge function. Aerosp. Sci. Technol. **25**(1), 177–189 (2013)
23. Hastie, T., Tibshirani, R., Friedman, J., Franklin, J.: The elements of statistical learning: data mining, inference and prediction. Math. Intell. **27**(2), 83–85 (2005)
24. Hensman, J., Fusi, N., Lawrence, N.D.,Gaussian processes for big data. arXiv preprint arXiv: 1309.6835 (2013)
25. Higdon, D., Gattiker, J., Williams, B., Rightley, M.: Computer model calibration using high-dimensional output. J. Am. Stat. Assoc. **103**(482), 570–583 (2008)
26. Huang, Z., Wang, C., Chen, J., Tian, H.: Optimal design of aeroengine turbine disc based on kriging surrogate models. Comput. Struct. **89**(1), 27–37 (2011)
27. Kennedy, M.C., O'Hagan, A.: Predicting the output from a complex computer code when fast approximations are available. Biometrika **87**(1), 1–13 (2000)

28. Koziel, S., Bekasiewicz, A., Couckuyt, I., Dhaene, T.: Efficient multi-objective simulation-driven antenna design using co-kriging. IEEE Trans. Antennas Propag. **62**(11), 5900–5905 (2014)
29. Madsen, J.I., Langthjem, M.: Multifidelity response surface approximations for the optimum design of diffuser flows. Optim. Eng. **2**(4), 453–468 (2001)
30. Mohan, S.C., Maiti, D.K.: Structural optimization of rotating disk using response surface equation and genetic algorithm. Int. J. Comput. Methods Eng. Sci. Mech. **14**(2), 124–132 (2013)
31. Neal, R.M.: Monte carlo implementation of Gaussian process models for Bayesian regression and classification. arXiv preprint physics/9701026 (1997)
32. Park, J.-S.: Optimal Latin-hypercube designs for computer experiments. J. Stat. Plann. Infer. **39**(1), 95–111 (1994)
33. Park, S., Choi, S.: Hierarchical Gaussian process regression. In: ACML, pp. 95–110 (2010)
34. Pepelyshev, A.: The role of the nugget term in the Gaussian process method. In: Giovagnoli, A., Atkinson, A.C., Torsney, B., May, C. (eds.) mODa 9-Advances in Model-Oriented Design and Analysis, pp. 149–156. Springer, Heidelberg (2010)
35. Qian, Z., Seepersad, C.C., Joseph, V.R., Allen, J.K., Wu, C.F.: Building surrogate models based on detailed and approximate simulations. J. Mech. Des. **128**(4), 668–677 (2006)
36. Quiñonero-Candela, J., Rasmussen, C.E.: A unifying view of sparse approximate Gaussian process regression. J. Mach. Learn. Res. **6**, 1939–1959 (2005)
37. Rasmussen, C.E., Williams, C.K.I.: Gaussian processes for machine learning. The MIT Press, Cambridge (2006)
38. Shaby, B., Ruppert, D.: Tapered covariance: Bayesian estimation and asymptotics. J. Comput. Graphical Stat. **21**(2), 433–452 (2012)
39. Shi, J.Q., Murray-Smith, R., Titterington, D.M.: Hierarchical Gaussian process mixtures for regression. Stat. Comput. **15**(1), 31–41 (2005)
40. Sun, G., Li, G., Stone, M., Li, Q.: A two-stage multi-fidelity optimization procedure for honeycomb-type cellular materials. Comput. Mater. Sci. **49**(3), 500–511 (2010)
41. Sun, S., Zhao, J., Zhu, J.: A review of Nyström methods for large-scale machine learning. Inf. Fusion **26**, 36–48 (2015)
42. Titsias, M.K.: Variational learning of inducing variables in sparse Gaussian processes. In: International Conference on Artificial Intelligence and Statistics, pp. 567–574 (2009)
43. Williams, C.K.I., Barber, D.: Bayesian classification with Gaussian processes. IEEE Trans. Pattern Anal. Mach. Intell. **20**(12), 1342–1351 (1998)
44. Xu, W., Tran, T., Srivastava, R., Journel, A.: Integrating seismic data in reservoir modeling: the collocated cokriging alternative. Society of Petroleum Engineers, In: SPE Annual Technical Conference and Exhibition (1992)
45. Zahir, M.K., Gao, Z.: Variable fidelity surrogate assisted optimization using a suite of low fidelity solvers. Open J. Optim. **1**(1), 0–8 (2012)
46. Zaitsev, A., Burnaev, E., Spokoiny, V.: Properties of the posterior distribution of a regression model based on Gaussian random fields. Autom. Remote Control **74**(10), 1645–1655 (2013)

Effective Design for Sobol Indices Estimation Based on Polynomial Chaos Expansions

Evgeny Burnaev[1], Ivan Panin[1(✉)], and Bruno Sudret[2]

[1] Kharkevich Institute for Information Transmission Problems,
Bolshoy Karetny per. 19, Moscow 127994, Russia
burnaev@iitp.ru, panin@phystech.edu
[2] Chair of Risk, Safety and Uncertainty Quantification, ETH Zurich,
Stefano-Franscini-Platz 5, 8093 Zurich, Switzerland
sudret@ibk.baug.ethz.ch

Abstract. Sobol' indices are a common metric of dependency in sensitivity analysis. It is used as a measure of confidence of input variables influence on the output of the analyzed mathematical model. We consider a problem of selection of experimental design points for Sobol' indices estimation. Based on the concept of D-optimality, we propose a method for constructing an adaptive design of experiments, effective for the calculation of Sobol' indices from Polynomial Chaos Expansions. We provide a set of applications that demonstrate the efficiency of the proposed approach.

Keywords: Design of experiment · Sensitivity analysis · Sobol indices · Polynomial chaos expansions · Active learning

1 Introduction

Computational models play an important role in different areas of human activity [1–3]. Over the past decades, they have become more complex, and there is an increasing need for special methods for the analysis of computational models. *Sensitivity analysis* is an important tool for investigation of computational models.

Sensitivity analysis tries to find how different model input parameters influence the model output, what are the most influential parameters and how to evaluate such effects quantitatively [4]. Sensitivity analysis allows to better understand the behavior of computational models. Particularly, it allows us to separate all input parameters into *important (significant)*, *relatively important* and *unimportant (nonsignificant)* ones. Important parameters, *i.e.* parameters whose variability has a strong effect on the model output, need to be controlled more accurately. Complex computational models often suffer from over-parameterization. By excluding unimportant parameters, we can potentially improve model quality, reduce parametrization (which is of great interest in the field of meta-modeling) and computational costs [26].

© Springer International Publishing Switzerland 2016
A. Gammerman et al. (Eds.): COPA 2016, LNAI 9653, pp. 165–184, 2016.
DOI: 10.1007/978-3-319-33395-3_12

Sensitivity analysis includes a wide range of metrics and techniques *e.g.* the Morris method [5], linear regression-based methods [6], variance-based methods [7]. Among others, *Sobol' (sensitivity) indices* are a common metric to evaluate the influence of model parameters [10]. Sobol' indices describe the portion of the output variance explained by different input parameters and combinations thereof. This method is especially useful for the case of nonlinear computational models [11].

There are two main approaches to evaluation of Sobol' indices. *Monte Carlo approach* (Monte Carlo simulation, FAST [12], SPF scheme [13] and others) is relatively robust [8], but requires large number of model runs, typically in the order of 10^4 for an accurate estimation of each index. Thus, it is impractical for a number of industrial applications, where each model evaluation is computationally costly.

Metamodeling approaches for Sobol' indices estimation allows one to reduce the necessary number of model runs [6,9]. Following this approach, we replace the original computational model by an approximating *metamodel* (also known as *surrogate model* or *response surface*) which is computationally efficient and has a clear internal structure [26]. The approach consists of the following general steps: selection of *the design of experiment (DoE)* and generation of *the training sample set*, construction of the metamodel based on the training samples, including quality assessment and evaluation of Sobol' indices (or any other measure) using the constructed metamodel. Note that the evaluation of indices may be either based on known internal structure of the metamodel or via Monte Carlo simulation on the metamodel itself.

In general, a metamodeling approach is more computationally efficient than a crude Monte Carlo approach, since the cost (in terms of the number of runs of the costly computational model) reduces to that of the training set (in general, a few dozens to a few hundreds). However, this approach may be nonrobust and its accuracy is more difficult for analysis. Indeed, although procedures like *cross-validation* [14,26] allow to estimate the quality of metamodels, the accuracy of complex statistics (*e.g.* Sobol' indices), derived from metamodels, has a complicated dependency on the metamodels structure and quality (see *e.g.* confidence intervals for Sobol' indices estimates [15] in case of Gaussian Processes metamodel [16–18] and bootstrap-based confidence intervals in case of polynomial chaos expansions [19]).

In this paper, we consider a problem of a design construction for a particular metamodeling approach: how to select the experimental design for building a polynomial chaos expansion for further evaluation of Sobol' indices, that is effective in terms of the number of computational model runs?

Space-filling designs are commonly used for sensitivity analysis. Methods like Monte Carlo sampling, *Latin Hypercube Sampling (LHS)* [20] or sampling in FAST method [12] try to fill "uniformly" the input parameters space with *design points* (*point*s are some realizations of parameters values). These sampling methods are *model free*, as they make no assumptions on the computational model.

In order to speed up the convergence of indices estimates, we assume that the computational model is close to its approximating metamodel and exploit knowledge of the metamodel structure. In this paper, we consider *Polynomial Chaos Expansions (PCE)* as a metamodel that is commonly used in engineering and other applications [21]. PCE approximation is based on a series of polynomials (Hermite, Legendre, Laguerre etc.) that are orthogonal w.r.t. the probability distributions of corresponding input parameters of the computational model. It allows to calculate Sobol' indices analytically from the expansion coefficients [22,23].

In this paper, we address the problem of design of experiments construction for evaluating Sobol' indices from a PCE metamodel. Based on asymptotic considerations, we propose an adaptive algorithm for design and test it on a set of applied problems. Note that in [33], we investigated the adaptive design algorithm for the case of a quadratic metamodel (see also [34]). In this paper, we extend the results for the case of a generalized PCE metamodel and provide more examples, including real industrial applications.

The paper is organized as follows: in Sect. 2, we review the definition of sensitivity indices and describe their estimation based on PCE metamodels. In Sect. 3, asymptotic behavior of indices estimates is obtained. In Sect. 4, we introduce an optimality criterion and propose a procedure for constructing the experimental design. In Sect. 5, we provide experimental results, applications and benchmark with other methods of design construction.

2   Sensitivity Indices and PCE Metamodels

2.1   Sensitivity Indices

Consider a computational model $y = f(\mathbf{x})$, where $\mathbf{x} = (x_1, \ldots, x_d) \in \mathscr{X} \subset \mathbb{R}^d$ is a vector of *input variables* (or *parameters* or *features*), $y \in \mathbb{R}^1$ is an *output variable* and $\mathscr{X}$ is *a design space*. The model $f(\mathbf{x})$ describes the behavior of some physical system of interest.

We consider the model $f(\mathbf{x})$ as a black-box: no additional knowledge on its inner structure is assumed. For the selected *design of experiments* $X = \{\mathbf{x}_i \in \mathscr{X}\}_{i=1}^n \in \mathbb{R}^{n \times d}$ we can obtain a set of model responses and form *a training sample*

$$L = \{\mathbf{x}_i, y_i = f(\mathbf{x}_i)\}_{i=1}^n \triangleq \{X \in \mathbb{R}^{n \times d}, Y = f(X) \in \mathbb{R}^n\}, \tag{1}$$

which allows us to investigate properties of the computational model.

Assume there is a prescribed probability distribution $\mathscr{H}$ with independent marginal distributions on the design space $\mathscr{X}$ ($\mathscr{H} = \mathscr{H}_1 \times \ldots \times \mathscr{H}_d$). This distribution represents the uncertainty and/or variability of the input variables, modelled as a random vector $\boldsymbol{X} = \{X_1, \ldots, X_d\}$ with independent components. In these settings, the model output becomes stochastic variable $\boldsymbol{Y} = f(\boldsymbol{X})$.

Assuming that the function $f(\boldsymbol{X})$ is square-integrable with respect to distribution $\mathscr{H}$ (*i.e.* $\mathbb{E}[f^2(\boldsymbol{X})] < +\infty$)), we have the following unique Sobol' decomposition of $\boldsymbol{Y} = f(\boldsymbol{X})$ [10] given by

$$f(\boldsymbol{X}) = f_0 + \sum_{i=1}^{d} f_i(X_i) + \sum_{1 \le i \le j \le d} f_{ij}(X_i, X_j) + \ldots + f_{1\ldots d}(X_1, \ldots, X_d),$$

which satisfies:
$$\mathbb{E}[f_{\mathbf{u}}(\boldsymbol{X}_{\mathbf{u}}) f_{\mathbf{v}}(\boldsymbol{X}_{\mathbf{v}})] = 0, \text{ if } \mathbf{u} \ne \mathbf{v},$$

where $\mathbf{u}$ and $\mathbf{v}$ are index sets: $\mathbf{u}, \mathbf{v} \subset \{1, 2, \ldots, d\}$.

Due to orthogonality of the summands, we can decompose the variance of the model output:

$$D = \mathbb{V}[f(\boldsymbol{X})] = \sum_{\substack{\mathbf{u} \subset \{1,\ldots,d\}, \\ \mathbf{u} \ne 0}} \mathbb{V}[f_{\mathbf{u}}(\boldsymbol{X}_{\mathbf{u}})] = \sum_{\substack{\mathbf{u} \subset \{1,\ldots,d\}, \\ \mathbf{u} \ne 0}} D_{\mathbf{u}},$$

In this expansion $D_{\mathbf{u}} \triangleq \mathbb{V}[f_{\mathbf{u}}(\boldsymbol{X}_{\mathbf{u}})]$ is the contribution of summand $f_{\mathbf{u}}(\boldsymbol{X}_{\mathbf{u}})$ to the output variance, also known as *partial variance*.

Definition 1. *The sensitivity index (Sobol' index) of variable set* $\boldsymbol{X}_{\mathbf{u}}$, $\mathbf{u} \subset \{1, \ldots, d\}$ *is defined as*

$$S_{\mathbf{u}} = \frac{D_{\mathbf{u}}}{D}. \tag{2}$$

The sensitivity index describes the amount of the total variance explained by the uncertainties in the subset of model input variables $\boldsymbol{X}_{\mathbf{u}}$.

Remark 1. In this paper, we consider only sensitivity indices of type $S_i \triangleq S_{\{i\}}, i = 1, \ldots, d$, called *first-order* or *main effect sensitivity indices*.

2.2 Polynomial Chaos Expansions

Consider a set of multivariate polynomials $\{\Psi_{\boldsymbol{\alpha}}(\boldsymbol{X}), \ \boldsymbol{\alpha} \in \mathscr{L}\}$ that consists of polynomials $\Psi_{\boldsymbol{\alpha}}$ having the form of tensor product

$$\Psi_{\boldsymbol{\alpha}}(\boldsymbol{X}) = \prod_{i=1}^{d} \psi_{\alpha_i}^{(i)}(X_i), \ \boldsymbol{\alpha} = \{\alpha_i \in \mathbb{N}, \ i = 1, \ldots, d\} \in \mathscr{L},$$

where $\psi_{\alpha_i}^{(i)}$ is a univariate polynomial of degree α_i belonging i-th family (*e.g.* Legendre polynomials, Jacobi polynomials, etc.), $\mathbb{N} = \{0, 1, 2, \ldots\}$ is the set of nonnegative integers, $\mathscr{L}$ is some fixed set of multi-indices $\boldsymbol{\alpha}$.

Suppose that univariate polynomials $\{\psi_{\alpha}^{(i)}\}$ are orthogonal w.r.t. i-th marginal of the probability distribution $\mathscr{H}$, *i.e.* $\mathbb{E}[\psi_{\alpha}^{(i)}(X_i)\psi_{\beta}^{(i)}(X_i)] = 0$ if $\alpha \ne \beta$ for $i = 1, \ldots, d$. Particularly, Legendre polynomials are orthogonal w.r.t. standard uniform distribution; Hermite polynomials are orthogonal w.r.t. Gaussian distribution. Due to independence of $\boldsymbol{X}$'s components, we obtain that multivariate polynomials $\{\Psi_{\boldsymbol{\alpha}}\}$ are orthogonal w.r.t. the probability distribution $\mathscr{H}$ *i.e.* $\mathbb{E}[\Psi_{\boldsymbol{\alpha}}(\boldsymbol{X})\Psi_{\boldsymbol{\beta}}(\boldsymbol{X})] = 0$ if $\boldsymbol{\alpha} \ne \boldsymbol{\beta}$.

Provided $\mathbb{E}[f^2(\boldsymbol{X})] < +\infty$, the spectral polynomial chaos expansion of f reads

$$f(\boldsymbol{X}) = \sum_{\boldsymbol{\alpha} \in \mathbb{N}^d} c_{\boldsymbol{\alpha}} \Psi_{\boldsymbol{\alpha}}(\boldsymbol{X}), \tag{3}$$

where $\{c_{\boldsymbol{\alpha}}\}$ are coefficients.

In the sequel we consider a PCE approximation $f_{PC}(\boldsymbol{X})$ of the model $f(\boldsymbol{X})$ obtained by truncating the infinite series to a finite number of terms:

$$\hat{y} = f_{PC}(\boldsymbol{X}) = \sum_{\boldsymbol{\alpha} \in \mathscr{L}} c_{\boldsymbol{\alpha}} \Psi_{\boldsymbol{\alpha}}(\boldsymbol{X}). \tag{4}$$

By enumerating the elements of $\mathscr{L}$ we will also use the alternative form of (4):

$$\hat{y} = f_{PC}(\boldsymbol{X}) = \sum_{\boldsymbol{\alpha} \in \mathscr{L}} c_{\boldsymbol{\alpha}} \Psi_{\boldsymbol{\alpha}}(\boldsymbol{X}) \triangleq \sum_{j=0}^{P-1} c_j \Psi_j(\boldsymbol{X}) = \mathbf{c}^T \boldsymbol{\Psi}(\boldsymbol{X}), \;\; P \triangleq |\mathscr{L}|,$$

where $\mathbf{c} = (c_0, \ldots, c_{P-1})^T$ is a column vector of coefficients and $\boldsymbol{\Psi}(\mathbf{x}) \colon \mathbb{R}^d \to \mathbb{R}^P$ is a mapping from the design space to *the extended design space* defined as a column vector function $\boldsymbol{\Psi}(\mathbf{x}) = (\Psi_0(\mathbf{x}), \ldots, \Psi_{P-1}(\mathbf{x}))^T$. Note that index $j = 0$ corresponds to multi-index $\boldsymbol{\alpha} = \mathbf{0} = \{0, \ldots, 0\}$ *i.e.*

$$c_{j=0} \triangleq c_{\boldsymbol{\alpha}=\mathbf{0}}, \; \Psi_{j=0} \triangleq \Psi_{\boldsymbol{\alpha}=\mathbf{0}} = const.$$

The set of multi-indices $\mathscr{L}$ is determined by *the truncation scheme*. In this work, we use hyperbolic truncation scheme [24], which corresponds to

$$\mathscr{L} = \{\boldsymbol{\alpha} \in \mathbb{N}^d : \|\boldsymbol{\alpha}\|_q \le p\}, \; \|\boldsymbol{\alpha}\|_q \triangleq \left(\sum_{i=1}^d \alpha_i^q\right)^{1/q},$$

where $q \in (0, 1]$ is a fixed parameter and $p \in \mathbb{N}\backslash\{0\} = \{1, 2, 3, \ldots\}$ is a fixed maximal total degree of polynomials. Note that in case of $q = 1$, we have $P = \frac{(d+p)!}{d!p!}$ polynomials in $\mathscr{L}$ and smaller q leads to a smaller number of polynomials.

There is a number of strategies for calculating the expansion coefficients $c_{\boldsymbol{\alpha}}$ in (4). In this paper, *the least-square (LS)* minimization method is used [25]. Expansion coefficients are calculated via minimization of the approximation error on the training sample $L = \{\mathbf{x}_i, y_i = f(\mathbf{x}_i)\}_{i=1}^n$:

$$\hat{\mathbf{c}}_{LS} = \arg \min_{\mathbf{c} \in \mathbb{R}^P} \frac{1}{n} \sum_{i=1}^n \left[y_i - \mathbf{c}^T \boldsymbol{\Psi}(\mathbf{x}_i)\right]^2. \tag{5}$$

2.3   PCE Post-processing for Sensitivity Analysis

Consider a PCE model $f_{PC}(\boldsymbol{X}) = \sum_{\boldsymbol{\alpha} \in \mathscr{L}} c_{\boldsymbol{\alpha}} \Psi_{\boldsymbol{\alpha}}(\boldsymbol{X}) = \sum_{j=0}^{P-1} c_j \Psi_j(\boldsymbol{X})$. According to [22], we have an explicit form of Sobol' indices (main effects) for model $f_{PC}(\boldsymbol{X})$:

$$S_i(\mathbf{c}) = \frac{\sum_{\boldsymbol{\alpha} \in \mathscr{L}_i} c_{\boldsymbol{\alpha}}^2 \mathbb{E}[\Psi_{\boldsymbol{\alpha}}^2(\boldsymbol{X})]}{\sum_{\boldsymbol{\alpha} \in \mathscr{L}_*} c_{\boldsymbol{\alpha}}^2 \mathbb{E}[\Psi_{\boldsymbol{\alpha}}^2(\boldsymbol{X})]}, \; i = 1, \ldots, d, \tag{6}$$

where $\mathscr{L}_* \triangleq \mathscr{L}\backslash\{0\}$ and $\mathscr{L}_i \subset \mathscr{L}$ is the set of multi-indices α such that only the index on the i-th position is nonzero: $\alpha = \{0, \ldots, \alpha_i, \ldots, 0\}$, $\alpha_i \in \mathbb{N}$, $\alpha_i > 0$.

Suppose for simplicity that the multivariate polynomials $\{\Psi_\alpha(X), \ \alpha \in \mathscr{L}\}$ are not only orthogonal but also normalized w.r.t. distribution $\mathscr{H}$:

$$\mathbb{E}[\Psi_\alpha(X)\Psi_\beta(X)] = \delta_{\alpha\beta}, \tag{7}$$

where $\delta_{\alpha\beta}$ is the Kronecker symbol, $i.e.$ $\delta_{\alpha\beta} = 1$ if $\alpha = \beta$, otherwise $\delta_{\alpha\beta} = 0$. Then (6) takes the form

$$S_i(\mathbf{c}) = \frac{\sum_{\alpha\in\mathscr{L}_i} c_\alpha^2}{\sum_{\alpha\in\mathscr{L}_*} c_\alpha^2}, \ i = 1, \ldots, d. \tag{8}$$

Thus, (8) gives a simple expression for the calculation of Sobol' indices for PCE metamodels. If the original model of interest $f(X)$ is close to its PCE approximation $f_{PC}(X)$, then we can use the expression for indices (8) with estimated coefficients (5) to approximate Sobol' indices of the original model:

$$\hat{S}_i = S_i(\hat{\mathbf{c}}) = \frac{\sum_{\alpha\in\mathscr{L}_i} \hat{c}_\alpha^2}{\sum_{\alpha\in\mathscr{L}_*} \hat{c}_\alpha^2}, \ i = 1, \ldots, d, \tag{9}$$

where $\hat{\mathbf{c}} \triangleq \hat{\mathbf{c}}_{LS}$.

3   Asymptotic Properties

In this section, we consider asymptotic properties of the indices estimates in Eq. (9) if the coefficients (5) are estimated on noisy data. Unlike (3), the original model is supposed to be the truncated PC with Gaussian noise:

$$f(\mathbf{x}) = f_{PC}(\mathbf{x}) + \varepsilon = \sum_{j=0}^{P-1} c_j \Psi_j(\mathbf{x}) + \varepsilon = \mathbf{c}^T \boldsymbol{\Psi}(\mathbf{x}) + \varepsilon, \tag{10}$$

where $\varepsilon \sim N(0, \sigma^2)$ is i.i.d. Gaussian noise. Let $\hat{\mathbf{c}}_n$ be the LS estimate (5) of the true coefficients vector $\mathbf{c}$ based on the training sample $L = \{\mathbf{x}_i, y_i = f(\mathbf{x}_i)\}_{i=1}^n$. In this section and further, if some variable has index n, then this variable depends on the training sample (1) of size n; and all expectations and variances are w.r.t. Gaussian noise.

Define *information matrix* $A_n \in \mathbb{R}^{P \times P}$ as

$$A_n = \sum_{i=1}^n \boldsymbol{\Psi}(\mathbf{x}_i)\boldsymbol{\Psi}^T(\mathbf{x}_i), \tag{11}$$

and assume that A_n is not degenerate, $i.e.$ $\det A_n \neq 0$. Then, using standard results for linear regression coefficients [26,27], we have

$$\mathbb{E}_\varepsilon[\hat{\mathbf{c}}_n] = \mathbf{c}, \quad \mathbb{V}_\varepsilon[\hat{\mathbf{c}}_n] = \sigma^2 A_n^{-1}$$

and

$$\hat{\mathbf{c}}_n - \mathbf{c} \sim \mathcal{N}(0, \, \sigma^2 A_n^{-1}). \tag{12}$$

The following theorem allows to establish asymptotic properties of the indices estimate based on the model (10) while new examples are added to the training sample.

Theorem 1. *1. Let us assume that there is an infinite sequence of points in the design space $\{\mathbf{x}_i \in \mathscr{X}\}_{i=1}^{\infty}$, such that*

$$\frac{1}{n} A_n = \frac{1}{n} \sum_{i=1}^{n} \boldsymbol{\Psi}(\mathbf{x}_i) \boldsymbol{\Psi}^T(\mathbf{x}_i) \xrightarrow[n \to +\infty]{} \Sigma, \tag{13}$$

where $\Sigma \in \mathbb{R}^{P \times P}$: $\Sigma = \Sigma^T$, $\det \Sigma > 0$, and new design points be added successively from this sequence to the design of experiments $X_n = \{\mathbf{x}_i\}_{i=1}^{n}$.
2. Let the sensitivity indices vector-function be defined by its components (8):

$$\mathbf{S}(\boldsymbol{\nu}) = (S_1(\boldsymbol{\nu}), \ldots, S_d(\boldsymbol{\nu}))^T$$

and $\hat{\mathbf{S}}_n \triangleq \mathbf{S}(\hat{\mathbf{c}}_n)$, where $\hat{\mathbf{c}}_n$ is defined by (5).
3. Assume that for the true coefficients $\mathbf{c}$ of the model (10):

$$\det(B\Sigma^{-1}B^T) \neq 0, \tag{14}$$

where B is the matrix of partial derivatives defined as

$$B \triangleq B(\mathbf{c}) = \left. \frac{\partial \mathbf{S}(\boldsymbol{\nu})}{\partial \boldsymbol{\nu}} \right|_{\boldsymbol{\nu}=\mathbf{c}} \in \mathbb{R}^{d \times P} \tag{15}$$

then

$$\sqrt{n} \left(\mathbf{S}(\hat{\mathbf{c}}_n) - \mathbf{S}(\mathbf{c}) \right) \xrightarrow[n \to +\infty]{\mathcal{D}} \mathcal{N}(0, \sigma^2 B\Sigma^{-1}B^T). \tag{16}$$

Proof

1. Condition (13) implies that $\det A_n \neq 0$ starting from some n_0. Therefore, (12) holds true for $n \geq n_0$. Now, consider only $n \geq n_0$.
2. From (12) and (13) we have

$$\sqrt{n}(\hat{\mathbf{c}}_n - \mathbf{c}) \sim \mathcal{N}(0, \, \sigma^2 (A_n/n)^{-1}) \xrightarrow[n \to +\infty]{\mathcal{D}} \mathcal{N}(0, \, \sigma^2 \Sigma^{-1}).$$

3. Applying δ-method [28] on vector-function $\mathbf{S}(\boldsymbol{\nu})$ at point $\boldsymbol{\nu} = \mathbf{c}$, we obtain required expression (16).

Remark 2. Note that the elements of B have the following form

$$b_{i\beta} \triangleq \frac{\partial S_i}{\partial c_\beta} = \begin{cases} \dfrac{2c_\beta \sum_{\alpha \in \mathscr{L}_*} c_\alpha^2 - 2c_\beta \sum_{\alpha \in \mathscr{L}_i} c_\alpha^2}{\left(\sum_{\alpha \in \mathscr{L}_*} c_\alpha^2\right)^2}, & \text{if } \beta \in \mathscr{L}_i, \\ 0, & \text{if } \beta = \mathbf{0} \triangleq \{0, \ldots, 0\}, \\ \dfrac{-2c_\beta \sum_{\alpha \in \mathscr{L}_i} c_\alpha^2}{\left(\sum_{\alpha \in \mathscr{L}_*} c_\alpha^2\right)^2}, & \text{if } \beta \notin \mathscr{L}_i \cup \mathbf{0}, \end{cases} \tag{17}$$

where $i = 1, \ldots, d$ and multi-index $\beta \in \mathscr{L}$. The elements of B can be also represented as

$$b_{i\beta} \triangleq \frac{\partial S_i}{\partial c_\beta} = \frac{-2c_\beta}{\sum_{\alpha \in \mathscr{L}_*} c_\alpha^2} \times \begin{cases} S_i - 1, & \text{if } \beta \in \mathscr{L}_i, \\ 0, & \text{if } \beta = \mathbf{0} \triangleq \{0, \ldots, 0\}, \\ S_i, & \text{if } \beta \notin \mathscr{L}_i \cup \mathbf{0}, \end{cases} \quad (18)$$

Remark 3. We can see that the theorem conditions do not depend on the type of orthonormal polynomials.

4   Design of Experiments Construction

4.1   Preliminary Considerations

Taking into account the results of Theorem 1, the limiting covariance matrix of the indices estimates depends on

1. Noise variance σ^2,
2. True values of PC coefficients $\mathbf{c}$, defining B,
3. Experimental design X, defining Σ.

  If we have a sufficiently accurate approximation of the original model, then in the above assumptions, the asymptotic formula (16) allows to evaluate the quality of the experimental design. Indeed, generally speaking the smaller the norm of the covariance matrix $\|\sigma^2 B \Sigma^{-1} B^T\|$, the better the estimation of the sensitivity indices. Theoretically, we could use this formula for constructing an experimental design that is effective for calculating Sobol' indices: we could select designs that minimize the norm of the covariance matrix. However, there are some problems when proceeding this way:

- The first one relates to the selection of the specific functional for minimization. Informally speaking, we need to choose "the norm" associated with the limiting covariance matrix.
- The second one refers to the fact that we do not know the true values of the PC model coefficients, defining B; therefore, we will not be able to accurately evaluate the quality of the design.

  The first problem can be solved in different ways. A number of statistical criteria for design optimality (D-, I-optimality and others, see [29]) are known. Similar to the work [33], we use the D-optimality criterion, as it a provides computationally efficient procedure for design construction. D-optimal experimental design minimizes the determinant of the limiting covariance matrix. If the vector of the estimated parameters is normally distributed then D-optimal design allows to minimize the volume of the confidence region for this vector.

  The second problem is more complex. The optimal design for estimating sensitivity indices that minimizes the norm of limiting covariance matrix depends on the true values of the indices, so it can be constructed only if these true values are known. However, in this case design construction makes no sense.

The dependency of the optimal design for indices evaluation on the true model parameters is a consequence of the nonlinearity of the indices estimates w.r.t. the PC model coefficients. In order to underline this dependency, the term "*locally D*-optimal design" is commonly used [30]. In these settings, there are several approaches, which are usually associated with either some assumptions about the unknown parameters, or adaptive design construction [30]. We use the latter approach.

In the case of adaptive designs, new design points are generated sequentially based on current estimate of the unknown parameters. This allows to avoid a priori assumptions on these parameters. However, this approach has a problem with a confidence of the solution found: if at some step of design construction parameters estimates are significantly different from their true values, then the design, which is constructed based on these estimates, may lead to new parameters estimates, which are also different from the true values.

In practice, during the construction of adaptive design, the quality of the approximation model and assumptions on non-degeneracy of results can be checked at each iteration and one can control and adjust the adaptive strategy.

4.2 Adaptive Algorithm

In this section, we introduce the adaptive algorithm for constructing a design of experiments that is effective to estimate sensitivity indices based on the asymptotic D-optimality criterion (see the algorithm scheme and Fig. 1). As it was discussed, the main idea of the algorithm is to minimize the confidence region for indices estimates. At each iteration, we replace the limiting covariance matrix by its approximation based on the current PC coefficients estimates.

As for initialization, we suppose that there is some *initial design*, and we require that this initial design is non-degenerate, *i.e.* such that the initial information matrix A_0 is nonsingular ($\det A_0 \neq 0$). In addition, at each iteration the non-degeneracy of the matrix $B_i A_i^{-1} B_i^T$, related to the criterion to be minimized, is checked.

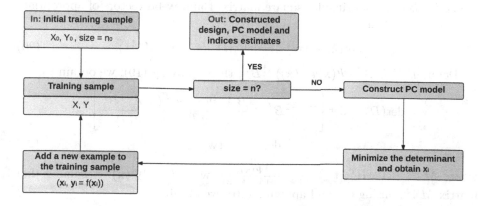

Fig. 1. Adaptive algorithm for constructing an effective experimental design to evaluate PC-based Sobol' indices

Goal: Construct an effective experimental design for the calculation of sensitivity
   indices
Parameters: initial and final numbers of points n_0 and n in the design; set of candidate
   design points $\varXi$.
Initialization:

- initial training sample $\{X_0, Y_0\}$ of size n_0, where design $X_0 = \{\mathbf{x}_i\}_{i=1}^{n_0} \subset \varXi$ defines
   a non-degenerate information matrix $A_0 = \sum_{i=1}^{n_0} \boldsymbol{\Psi}(\mathbf{x}_i)\boldsymbol{\Psi}^T(\mathbf{x}_i)$;
- $B_0 = B(\hat{\mathbf{c}}_0)$, obtained using the initial estimates of the coefficients of PC model,
   see (15), (17), (18);

Iterations: for all i from 1 to $n - n_0$:

- $\mathbf{x}_i = \arg\min_{\mathbf{x} \in \varXi} \det\left[B_{i-1}(A_{i-1} + \boldsymbol{\Psi}(\mathbf{x})\boldsymbol{\Psi}^T(\mathbf{x}))^{-1} B_{i-1}^T \right]$
- $A_i = A_{i-1} + \boldsymbol{\Psi}(\mathbf{x}_i)\boldsymbol{\Psi}^T(\mathbf{x}_i)$
- Add the new sample point $(\mathbf{x}_i, y_i = f(\mathbf{x}_i))$ to the training sample and update
   current estimates $\hat{\mathbf{c}}_i$ of the PCE model coefficients
- Calculate $B_i = B(\hat{\mathbf{c}}_i)$

Output: The design of experiment $X = X_0 \cup X_{add}$, where $X_{add} = \{\mathbf{x}_k\}_{k=1}^{n-n_0}$

4.3   Details of the Optimization Procedure

The idea behind the proposed optimization procedure is analogous to the idea
of the Fedorov's algorithm for constructing optimal designs [31].
   In order to simplify the optimization problem, we use two well-known iden-
tities:

- Let M be some nonsingular square matrix, $\mathbf{t}$ and $\mathbf{w}$ be vectors such that
   $1 + \mathbf{w}^T M^{-1}\mathbf{t} \neq 0$, then

$$(M + \mathbf{t}\mathbf{w}^T)^{-1} = M^{-1} - \frac{M^{-1}\mathbf{t}\mathbf{w}^T M^{-1}}{1 + \mathbf{w}^T M^{-1}\mathbf{t}}. \tag{19}$$

- Let M be some nonsingular square matrix, $\mathbf{t}$ and $\mathbf{w}$ be vectors of appropriate
   dimensions, then

$$\det(M + \mathbf{t}\mathbf{w}^T) = \det(M) \cdot (1 + \mathbf{w}^T M^{-1}\mathbf{t}). \tag{20}$$

Define $D \triangleq B(A + \boldsymbol{\Psi}(\mathbf{x})\boldsymbol{\Psi}^T(\mathbf{x}))^{-1}B^T$, then applying (19), we obtain

$$\det(D) = \det\left[BA^{-1}B^T - \frac{BA^{-1}\boldsymbol{\Psi}(\mathbf{x})\boldsymbol{\Psi}^T(\mathbf{x})A^{-1}B^T}{1 + \boldsymbol{\Psi}^T(\mathbf{x})A^{-1}\boldsymbol{\Psi}(\mathbf{x})} \right]$$
$$\triangleq \det\left[M - \mathbf{t}\mathbf{w}^T \right], \tag{21}$$

where $M \triangleq BA^{-1}B^T$, $\mathbf{t} \triangleq \frac{BA^{-1}\boldsymbol{\Psi}(\mathbf{x})}{1 + \boldsymbol{\Psi}^T(\mathbf{x})A^{-1}\boldsymbol{\Psi}(\mathbf{x})}$, $\mathbf{w} \triangleq BA^{-1}\boldsymbol{\Psi}(\mathbf{x})$. Assuming that
matrix M is nonsingular and applying (20), we obtain

$$\det(D) = \det(M) \cdot (1 - \mathbf{w}^T M^{-1}\mathbf{t}) \rightarrow \min$$

The resulting optimization problem is

$$\mathbf{w}^T M^{-1} \mathbf{t} \to \max \tag{22}$$

or explicitly

$$\frac{(\boldsymbol{\Psi}^T(\mathbf{x})A^{-1})B^T(BA^{-1}B^T)^{-1}B(A^{-1}\boldsymbol{\Psi}(\mathbf{x}))}{1 + \boldsymbol{\Psi}^T(\mathbf{x})A^{-1}\boldsymbol{\Psi}(\mathbf{x})} \to \max_{\mathbf{x} \in \Xi}. \tag{23}$$

5 Benchmark

In this section, we validate the proposed algorithm on a set of computational models of different input dimensions. Three analytic problems and two industrial problems based on finite element models are considered. Input parameters (variables) of the considered models have independent uniform and independent normal distributions. For some models, independent gaussian noise is added to their outputs.

At first, we form non-degenerate random initial design, and then we use various techniques to add new design points iteratively. We compare our method for design construction (denoted as **Adaptive for SI**) with the following methods:

- **Random** method iteratively adds new design points randomly from the set of candidate design points Ξ;
- **Adaptive D-opt** iteratively adds new design points that maximize the determinant of the information matrix (11): $\det A_n \to \max_{\mathbf{x}_n \in \Xi}$ [31]. The resulting design is optimal, in some sense, for estimation of the coefficients of a PCE model. We compare our method with this approach to prove that it gives some advantage over usual D-optimality. Strictly speaking, D-optimal design is not iterative but if we have an initial training sample then the sequential approach seems natural generalization of a common D-optimal designs.
- **LHS**. Unlike other considered designs, this method is not iterative as a completely new design is generated at each step. This method uses Latin Hypercube Sampling, as it is common to compute PCE coefficients.

The metric of design quality is **the mean error** defined as the distance between estimated and true indices $\sqrt{\sum_{i=1}^{d}(S_i - \hat{S}_i^{\mathrm{run}})^2}$ averaged over runs with different random initial designs (200–400 runs). We consider not only *mean* error but also its *variance*. Particularly, we use Welch's t-test [32] to ensure that the difference of mean distances is statistically significant for the various methods considered. Note that lower p-values correspond to greater confidence.

In all cases, we assume that the truncation set (retained PCE terms) was selected before the experiment.

5.1 Analytic Functions

Two analytic functions with uniformly distributed input variables are considered, namely Sobol and Wing Weight functions. Independent gaussian noise is added to their outputs to simulate random errors due to measurement uncertainty.

The Sobol' function is commonly used for benchmarking methods in global sensitivity analysis

$$f(\mathbf{x}) = \prod_{i=1}^{d} \frac{|4x_i - 2| + c_i}{1 + c_i},$$

where $x_i \sim \mathcal{U}(0,1)$. In our case, parameters $d = 3$, $c = (0.0, 1.0, 1.5)$ are used. Independent gaussian noise is added to the output of the function. The standard deviation of noise is 0 (without noise), 0.2 and 1.4 that corresponds to 0 %, 28 % and 194 % of the function standard deviation due to given uncertainty of the inputs. Analytical expressions for its sensitivity indices are available in [10].

The Wing Weight function models the weight of aircraft wing [35]

$$f(\mathbf{x}) = 0.036 S_w^{0.758} W_{fw}^{0.0035} \left(\frac{A}{\cos^2(\Lambda)}\right)^{0.6} q^{0.006} \lambda^{0.04} \left(\frac{100 t_c}{\cos(\Lambda)}\right)^{-0.3} (N_z W_{dg})^{0.49}$$
$$+ S_w W_p,$$

where 10 input variables and their distributions are defined as: $S_w \sim \mathcal{U}(150, 200)$, wing area (ft^2); $W_{fw} \sim \mathcal{U}(220, 300)$, weight of fuel in the wing (lb); $A \sim \mathcal{U}(6, 10)$, aspect ratio; $\Lambda \sim \mathcal{U}(-10, 10)$, quarter-chord sweep (degrees); $q \sim \mathcal{U}(16, 45)$, dynamic pressure at cruise (lb/ft^2); $\lambda \sim \mathcal{U}(0.5, 1)$, taper ratio; $t_c \sim \mathcal{U}(0.08, 0.18)$, aerofoil thickness to chord ratio; $N_z \sim \mathcal{U}(2.5, 6)$, ultimate load factor; $W_{dg} \sim \mathcal{U}(1700, 2500)$, flight design gross weight (lb); $W_p \sim \mathcal{U}(0.025, 0.08)$, paint weight (lb/ft^2). Independent gaussian noise $\mathcal{N}(0, 5.0^2)$ is added to the output of the function.

Experimental setup for analytic functions is summarized in Table 1. In the experiments, we assume that the set of candidate design points Ξ is a uniform grid in the d-dimensional hypercube. Note that Ξ affects the quality of optimization.

5.2   Finite Element Models

Case 1: Truss Model. The deterministic computational model, originating from [36], resembles the displacement V_1 of a truss structure with 23 members as shown in Fig. 2.

10 random variables are considered:

– E_1, E_2 (Pa) $\sim \mathcal{U}(1.68 \times 10^{11}, 2.52 \times 10^{11})$;
– A_1 $(m^2) \sim \mathcal{U}(1.6 \times 10^{-3}, 2.4 \times 10^{-3})$;
– A_2 $(m^2) \sim \mathcal{U}(0.8 \times 10^{-3}, 1.2 \times 10^{-3})$;
– P_1 - P_6 (N) $\sim \mathcal{U}(3.5 \times 10^4, 6.5 \times 10^4)$;

It is assumed that all the horizontal elements have perfectly correlated Young's modulus and cros-sectional areas with each other and so is the case with the diagonal members.

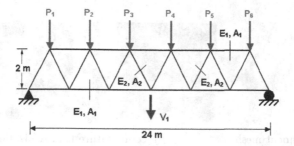

Fig. 2. Truss structure with 23 members

Case 2: Heat Transfer Model. We consider the two-dimensional stationary heat diffusion problem used in [37]. The problem is defined on the square domain $D = (-0.5, 0.5) \times (-0.5, 0.5)$ shown in Fig. 3a, where the temperature field $T(z), z \in D$, is described by the partial differential equation:

$$-\nabla(\kappa(\mathbf{z})\nabla T(z)) = 500 I_A(z),$$

with boundary conditions $T = 0$ on the top boundary and $\nabla T \mathbf{n} = 0$ on the left, right and bottom boundaries, where $\mathbf{n}$ denotes the vector normal to the boundary; $A = (0.2, 0.3) \times (0.2, 0.3)$ is a square domain within D and I_A is the indicator function of A. The diffusion coefficient, $\kappa(z)$, is a lognormal random field defined by

$$\kappa(z) = \exp[a_k + b_k g(z)],$$

where $g(z)$ is a standard Gaussian random field and the parameters a_k and b_k are such that the mean and standard deviation of κ are $\mu_\kappa = 1$ and $\sigma_\kappa = 0.3$, respectively. The random field $g(z)$ is characterized by an autocorrelation function $\rho(z, z') = \exp(-\|z - z'\|^2/0.2^2)$. The quantity of interest, Y, is the average temperature in the square domain $B = (-0.3, -0.2) \times (-0.3, -0.2)$ within D (see Fig. 3a).

To facilitate solution of the problem, the random field $g(z)$ is represented using the Expansion Optimal Linear Estimation (EOLE) method [38]. By truncating the EOLE series after the first M terms, $g(z)$ is approximated by

$$\hat{g}(z) = \sum_{i=1}^{M} \frac{\xi_i}{\sqrt{\ell_i}} \phi_i^T \mathbf{C}_{z\zeta}$$

In the above equation, $\{\xi_1, \ldots, \xi_M\}$ are independent standard normal variables; $\mathbf{C}_{z\zeta}$ is a vector with elements $\mathbf{C}_{z\zeta}^{(k)} = \rho(z, \zeta_k)$, where $\{\zeta_1, \ldots, \zeta_M\}$ are the points of an appropriately defined mesh in D; and (ℓ_i, ϕ_i) are the eigenvalues and eigenvectors of the correlation matrix $\mathbf{C}_{\zeta\zeta}$ with elements $\mathbf{C}_{\zeta\zeta}^{(k,\ell)} = \rho(\zeta_k, \zeta_\ell)$, where $k, \ell = 1, \ldots, n$. We select $M = 53$ in order to satisfy

$$\sum_{i=1}^{M} \ell_i / \sum_{i=1}^{n} \ell_i \geq 0.99$$

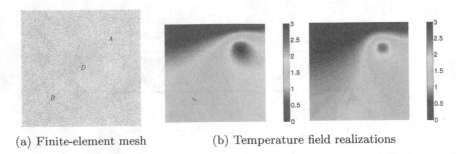

(a) Finite-element mesh   (b) Temperature field realizations

Fig. 3. Heat diffusion problem

The underlying deterministic problem is solved with an in-house finite-element analysis code. The employed finite-element discretization with triangular $T3$ elements is shown in Fig. 3a. Figure 3b shows the temperature fields corresponding to two example realizations of the diffusion coefficient.

Experimental Setup. For these finite element models, we assume that the set of candidate design points Ξ is

- a uniform grid in the 10-dimensional hypercube for Truss model;
- LHS design with normally distributed variables in 53-dimensional space for Heat transfer model.

Experimental settings for all models are summarized in Table 1.

Table 1. Benchmark settings

Characteristic	Sobol	WingWeight	Truss	Heat transfer
Input dimension	3	10	10	53
Input distributions	Unif	Unif	Unif	Norm
PCE degree	9	4	4	2
q-norm	0.75	0.75	0.75	0.75
Regressors number	111	176	176	107
Initial design size	150	186	176	108
Added noise std	(0, 0.2, 1.4)	5.0	—	—

5.3   Results

Figures 4a, b, c, and 5 show results for analytic functions. Figures 6 and 7 present results for finite element models.

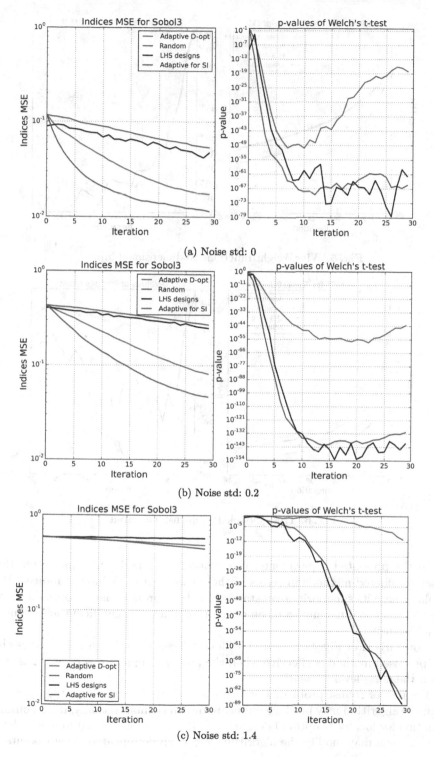

(a) Noise std: 0

(b) Noise std: 0.2

(c) Noise std: 1.4

Fig. 4. Sobol function. 3-dimensional input.

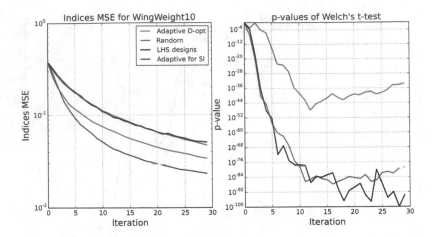

Fig. 5. WingWeight function. 10-dimensional input

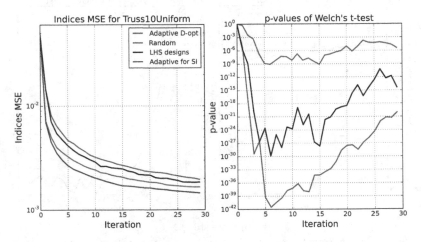

Fig. 6. Truss model. 10-dimensional input

In the presented experiments, the proposed method performs better than other considered methods in terms of the mean error of estimated indices. Particularly note its superiority over standard LHS approach that is commonly used in practice. The difference in mean errors is statistically significant according to Welch's t-test.

Comparison of Fig. 4a, b, c with different levels of additive noise shows that the proposed method is effective when the analyzed function is deterministic or when the noise level is *small*.

Because of robust problem statement and limited accuracy of the optimization, the algorithm may produce duplicate design points. Actually, it's a common situation for locally D-optimal designs [30]. If the computational model is deterministic, one may modify the algorithm, *e.g.* exclude repeated design points.

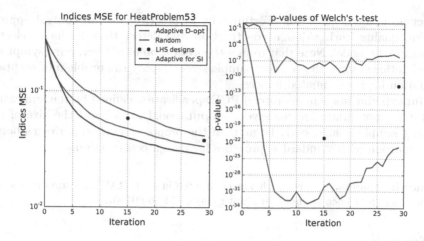

Fig. 7. Heat transfer model. 53-dimensional input

Although high dimensional optimization problems may be computationally prohibitive, the proposed approach is still useful in high dimensional settings. We propose to generate a uniform candidate set (*e.g.* LHS design of large size) and then choose its subset for the effective calculation of Sobol' indices using our adaptive method (see results for Heat transfer model on Fig. 7).

It should be noted that in all presented cases the specification of sufficiently accurate PCE model (reasonable values for degree and q-norm defining the truncation set) is assumed to be known *a priori* and the size of the initial training sample is sufficiently large. If we use an inadequate specification of the PCE model (*e.g.* quadratic PCE in case of cubic analyzed function), the method will perform worse in comparison with methods which do not depend on PCE model structure. In any case, usage of inadequate PCE models may lead to inaccurate results. That is why it is very important to control PCE model error during the design construction. For example, one may use *cross-validation* for this purpose [26]. Thus, if the PCE model error increases during design construction this may indicate that the model specification is inadequate and should be changed.

6   Conclusions

We proposed the design of experiments algorithm for evaluation of Sobol' indices from PCE metamodel. The method does not depend on a particular form of orthonormal polynomials in PCE. It can be used for the case of different distributions of input parameters of the analyzed computational models.

The main idea of the method comes from metamodeling approach. We assume that the computational model is close to its approximating PCE metamodel and exploit knowledge of a metamodel structure. This allows us to improve the evaluation accuracy. All comes with a price: if additional assumptions on the computational model to provide good performance are not satisfied, one may expect accuracy degradation. Fortunately, in practice, we can control approximation quality during design construction and detect that we have selected inappropriate model. Note that from a theoretical point of view, our asymptotic considerations (w.r.t. the training sample size) simplify the problem of accuracy evaluation for the estimated indices.

Our experiments demonstrate: if PCE specification defined by the truncation scheme is appropriate for the given computational model and the size of the training sample is sufficiently large, then the proposed method performs better in comparison with standard approaches for design construction.

Acknowledgments. The research was conducted in IITP RAS and supported solely by the Russian Science Foundation grant (project 14-50-00150).

References

1. Beven, K.J.: Rainfall-Runoff Modelling-The Primer, p. 360. Wiley, Chichester (2000)
2. Dayan, P., Abbott, L.F.: Theoretical Neuroscience: Computational and Mathematical Modeling of Neural Systems. MIT Press, Cambridge (2001)
3. Grihon, S., Burnaev, E.V., Belyaev, M.G., Prikhodko, P.V.: Surrogate modeling of stability constraints for optimization of composite structures. In: Koziel, S., Leifsson, L. (eds.) Surrogate-Based Modeling and Optimization. Engineering Applications, pp. 359–391. Springer, New York (2013)
4. Saltelli, A., Chan, K., Scott, M.: Sensitivity Analysis. Probability and Statistics Series. Wiley, West Sussex (2000)
5. Morris, M.D.: Factorial sampling plans for preliminary computational experiments. Technometrics **33**, 161–174 (1991)
6. Iooss, B., Lemaitre, P.: A review on global sensitivity analysis methods. In: Meloni, C., Dellino, G. (eds.) Uncertainty management in Simulation-Optimization of Complex Systems: Algorithms and Applications, pp. 101–122. Springer, New York (2010)
7. Saltelli, A., Ratto, M., Andres, T., Campolongo, F., Cariboni, J., Gatelli, D., et al.: Global Sensitivity Analysis - The Primer. Wiley, Chichester (2008)
8. Yang, J.: Convergence and uncertainty analyses in Monte-Carlo based sensitivity analysis. Environ. Modell. Softw. **26**, 444–457 (2011)
9. Sudret, B.: Polynomial chaos expansions and stochastic finite element methods. In: Phoon, K.K., Ching, J. (eds.) Risk and Reliability in Geotechnical Engineering, Chap. 6, pp. 265–300. Taylor and Francis, London (2015)
10. Sobol', I.M.: Sensitivity estimates for nonlinear mathematical models. Math. Model. Comp. Exp. **1**, 407–414 (1993)
11. Saltelli, A., Annoni, P.: How to avoid a perfunctory sensitivity analysis. Environ. Modell. Softw. **25**, 1508–1517 (2010)
12. Cukier, R.I., Levine, H.B., Shuler, K.E.: Nonlinear sensitivity analysis of multiparameter model systems. J. Comput. Phy. **26**(1), 1–42 (1978)
13. Sobol, I.: Global sensitivity indices for nonlinear mathematical models and their Monte Carlo estimates. Math. Comput. Simul. **55**(1–3), 271–280 (2001)
14. Stone, M.: Cross-validatory choice and assessment of statistical predictions. J. R. Stat. Soc., Ser. B **36**, 111–147 (1974)

15. Marrel, A., Iooss, B., Laurent, B., Roustant, O.: Calculations of the Sobol indices for the Gaussian process metamodel. Reliab. Eng. Syst. Saf. **94**, 742–751 (2009)
16. Burnaev, E., Zaitsev, A., Spokoiny, V.: Properties of the posterior distribution of a regression model based on Gaussian random fields. Autom. Remote Control **74**(10), 1645–1655 (2013)
17. Burnaev, E., Zaytsev, A., Spokoiny, V.: The Bernstein-von Mises theorem for regression based on Gaussian processes. Russ. Math. Surv. **68**(5), 954–956 (2013)
18. Belyaev, M., Burnaev, E., Kapushev, Y.: Gaussian process regression for structured data sets. In: Gammerman, A., Vovk, V., Papadopoulos, H. (eds.) SLDS 2015. LNCS, vol. 9047, pp. 106–115. Springer, Heidelberg (2015)
19. Dubreuila, S., Berveillerc, M., Petitjeanb, F., Salauna, M.: Construction of boot-strap confidence intervals on sensitivity indices computed by polynomial chaos expansion. Reliab. Eng. Syst. Saf. **121**, 263–275 (2014)
20. McKay, M.D., Beckman, R.J., Conover, W.J.: A comparison of three methods for selecting values of input variables in the analysis of output from a computer code. Technometrics **21**, 239–245 (1979)
21. Ghiocel, D., Ghanem, R.: Stochastic finite element analysis of seismic soil-structure interaction. J. Eng. Mech. **128**, 66–77 (2002)
22. Sudret, B.: Global sensitivity analysis using polynomial chaos expansions. Reliab. Eng. Sys. Safety **93**, 964–979 (2008)
23. Blatman, G., Sudret, B.: Efficient computation of global sensitivity indices using sparse polynomial chaos expansions. Reliab. Eng. Sys. Saf. **95**, 1216–1229 (2010)
24. Blatman, G., Sudret, B.: An adaptive algorithm to build up sparse polynomial chaos expansions for stochastic finite element analysis. Prob. Eng. Mech. **25**, 183–197 (2010b)
25. Berveiller, M., Sudret, B., Lemaire, M.: Stochastic finite elements: a non intrusive approach by regression. Eur. J. Comput. Mech. **15**(1–3), 81–92 (2006)
26. Hastie, T., Tibshirani, R., Friedman, J.: Elements of Statistical Learning: Data Mining, Inference and Prediction. Springer-Verlag, New York (2009)
27. Spokoiny, V., Dickhaus, T.: Basics of Modern Parametric Statistics. Springer, Berlin (2014)
28. Oehlert, G.W.: A note on the delta method. Am. Stat. **46**, 27–29 (1992)
29. Chaloner, K., Verdinelli, I.: Bayesian experimental design: a review. Stat. Sci. **10**, 273–304 (1995)
30. Pronzato, L.: One-step ahead adaptive D-optimal design on a finite design space is asymptotically optimal. Metrika **71**(2), 219–238 (2010)
31. Miller, A., Nguyen, N.K.: Algorithm AS 295: a fedorov exchange algorithm for D-optimal design. J. R. Stat. Soc. Ser. C (Appl. Stat. **43**(4), 669–677 (1994)
32. Welch, B.L.: The generalization of "Student's" problem when several different population variances are involved. Biometrika **34**(12), 2835 (1947)
33. Burnaev, E., Panin, I.: Adaptive Design of Experiments for Sobol Indices Estimation Based on Quadratic Metamodel. In: Gammerman, A., Vovk, V., Papadopoulos, H. (eds.) SLDS 2015. LNCS, vol. 9047, pp. 86–95. Springer, Heidelberg (2015)
34. Burnaev, E., Panov, M.: Adaptive design of experiments based on gaussian processes. In: Gammerman, A., Vovk, V., Papadopoulos, H. (eds.) SLDS 2015. LNCS, vol. 9047, pp. 116–125. Springer, Heidelberg (2015)
35. Forrester, A., Sobester, A., Keane, A.: Engineering Design via Surrogate Modelling: A Practical Guide. Wiley, Chichester (2008)
36. Lee, S., Kwak, B.: Response surface augmented moment method for efficient reliability analysis. Struct. Safe. **28**, 261–272 (2006)

37. Konakli, K., Sudret, B.: Uncertainty quantification in high-dimensional spaces with low-rank tensor approximations. In: Proceedings of the 1st ECCOMAS Thematic Conference on Uncertainty Quantification in Computational Sciences and Engineering, Crete Island, Greece (2015)
38. Li, C., Der Kiureghian, A.: Optimal discretization of random fields. J. Eng. Mech. **119**(6), 1136–1154 (1993)

Joint Prediction of Chronic Conditions Onset: Comparing Multivariate Probits with Multiclass Support Vector Machines

Shima Ghassem Pour[1,2](✉) and Federico Girosi[1,2]

[1] Centre for Health Research, Western Sydney University, Sydney, Australia
S.GhassemPour@WesternSydney.edu.au
[2] Capital Markets CRC, Sydney, Australia

Abstract. We consider the problem of building accurate models that can predict, in the short term (2–3 years), the onset of one or more chronic conditions at individual level. Five chronic conditions are considered: heart disease, stroke, diabetes, hypertension and cancer. Covariates for the models include standard demographic/socio-economic variables, risk factors and the presence of the chronic conditions at baseline. We compare two predictive models. The first model is the multivariate probit (MVP), chosen because it allows to model correlated outcome variables. The second model is the Multiclass Support Vector Machine (MSVM), a leading predictive method in machine learning. We use Australian data from the Social, Economic, and Environmental Factory (SEEF) study, a follow up to the 45 and Up Study survey, that contains two repeated observations of 60,000 individuals in NSW, over age 45. We find that MSVMs predictions have specificity rates similar to those of MVPs, but sensitivity rates that are on average 12 % points larger than those of MVPs, translating in a large average improvement in sensitivity of 30 %.

1 Introduction

While infectious disease continue to pose a threat to world health, in the words of the World Health Organization "it is the looming epidemics of heart disease, stroke, cancer and other chronic diseases that for the foreseeable future will take the greatest toll in deaths and disability"[1]. In fact, already 10 years ago the total number of people dying from chronic diseases was double that of all infectious diseases, maternal/perinatal conditions, and nutritional deficiencies combined [1]. The rise of these conditions can be traced to a complex web of interactions of common factors, such as genes, nutrition and life-style, with socio-economic status.

Since chronic conditions can be very costly but are also preventable there is great interest in building models that allow to simulate the costs and benefits of health interventions in this area, and that can be used for planning and policy purposes by government agencies and other interested stakeholders [2–4].

The risk of developing a chronic condition is highly dependent on factors such as obesity or smoking and on individual characteristics such as income

© Springer International Publishing Switzerland 2016
A. Gammerman et al. (Eds.): COPA 2016, LNAI 9653, pp. 185–195, 2016.
DOI: 10.1007/978-3-319-33395-3_13

and education. These factors vary greatly within the population, and therefore it is particular important to develop models that predict the onset of chronic conditions at individual level, that can then be used as components of simulation models to be applied to an entire population [2].

Since chronic conditions are quite correlated (for example diabetes and heart disease often go together) it is imperative to use models that make joint predictions, rather than modeling each condition separately. In the biostatistics literature this is usually done using multivariate probit models (MVP) [5,6]. While MVP are very attractive because they are easily interpretable, they rely on a very simple and rather restrictive specification and they were designed more for the purpose of understanding the determinants of the outcomes, rather than for predicting the future.

From a machine learning viewpoint it is somewhat surprising that there have not attempts to use more sophisticated, and appropriate, type of models, such as Support Vector Machines (SVMs) or Deep Learning (DL) methods. We start to fill this gap by presenting, in this paper, a comparison between the predictive ability of MVPs and SVMs. We have chosen SVMs to start with mostly because the biostatistics community is very comfortable with R and at the moment there is somewhat more support in R for SVMs than for DL.

It is important to underscore that the ability to improve the accuracy of MVP predictions is not an academic exercise. What is of interest to policy makers is long term predictions (20 to 30 years), that can only be made by repeatedly applying shorter term predictions (from one to three years, depending on the availability of longitudinal data). Therefore even a small improvement in the accuracy of short-term predictions can result in large reduction in the uncertainty of the long-term estimates, having a large impact on the policy outcomes.

The rest of this article is organized as follows. Section 2 describes the data used in our experiment. Section 3 briefly describes the MVP and SVM models. Section 4 discusses the experimental results and Sect. 5 concludes the paper.

2   Data

In order to build a predictive model of chronic disease it is necessary to have longitudinal data, in which the same individual has been observed at least twice. Since we are interested in predicting several chronic conditions at once, and since the joint prevalence of certain conditions is not very high, the data sets needs to be quite large in order to capture some of those combinations. There is a dearth of longitudinal data that can be used for this purpose, and one of the largest is the Australian Social, Economic, and Environmental Factory (SEEF) study, a follow up to the 45 and Up Study survey [7]. The approval for this study is provided by the NSW Population & Health Services Research Ethics Committee (AU RED reference:HREC/15/CIPHS/4).

The 45 and Up Study survey (www.saxinstitute.org.au), which was carried out between 2006 and 2009, contains information regarding the health and social wellbeing of 267,153 individuals aged 45 years and older living in New South

Wales (NSW), Australia. Eligible individuals, sampled from the Medicare population of NSW, were mailed the questionnaire, an information sheet and a consent form and provided with a reply paid envelope. The survey over-sampled individuals aged 80 years and over and residents of rural areas by a factor of two. In addition, all residents aged 45 years and older in remote areas were sampled. The overall response rate of the 45 and Up Study is 18%, accounting for approximately 10% of all individuals of age 45 years or older living in NSW. While the response rate is low and participants tended to be of more favorable socioeconomic circumstances than average for the age group, previous work has shown that analytical findings based on internal comparisons, such as odd-ratios, are generalizable and comparable to those derived from smaller but more representative population health surveillance [8].

Data captured in the 45 and Up Study baseline include a number of self-reported chronic conditions such as (ever diagnosed) heart disease, high blood pressure, diabetes, stroke, asthma, depression and different types of cancer.

Questionnaire data also include information on key potential confounder and mediating factors, including age, sex, household income, level of education, smoking history, alcohol use, physical activity, height and weight, functional status, psychological distress, medical and surgical history and dietary habits. A full description of all the variables available in the 45 and Up Study together with basic summary statistics can be found elsewhere [7].

The SEEF study data, that include all the original variable in the 45 and Up Study plus a host of additional variables, were collected in 2010 from a random sub sample of the baseline 45 and Up Study cohort. One hundred thousands 45 and Up Study participants were mailed an invitation and the SEEF questionnaire. About 60,000 individuals joined the SEEF study by completing the consent form and the questionnaire and mailing them to the study coordinating center.

Our dependent variables are 5 binary variables denoting the presence or absence of the following chronic conditions at follow-up: heart disease, hypertension, diabetes, stroke, and cancer. These health conditions were self-reported and based on the responses to survey questions formulated as follows: "Has a doctor ever told you that you have [name of condition]?".

Since individuals can develop any of those five conditions we consider the multi-class problem of predicting in which of the $2^5 = 32$ combinations of conditions individuals will fall at follow-up. We report in Table 1 the size of each of the 32 classes in the SEEF data. Since some of the classes are very small and neither of the two methods out-performed the other in those cases, we have eliminated from our data the classes with fewer than 100 cases (outlined in bold in Table 1).

The two main risk factors that we used as covariates were obesity and smoking status. Possible values of smoking status are "Not Smoking", "Smoker" and "quit smoking", which are derived from the combined answers to the following two questions "Have you ever been a regular smoker?" and "Are you a regular smoker now?".

Table 1. Class size (bold font shows the classes which we removed)

Condition	Size
No condition	16421
Cancer	11896
Cancer-hypertension	7614
Diabetes	737
Diabetes-Cancer	523
Diabetes-Cancer-Hypertension	1195
Diabetes-Hypertension	1389
Diabetes-Stroke	**19**
Diabetes-Stroke-Cancer	**20**
Diabetes-Stroke-Cancer-Hypertension	137
Diabetes-Stroke-Hypertension	**95**
Heart	1319
Heart-Cancer	1925
Heart-Cancer-Hypertension	2847
Heart-Diabetes	169
Heart-Diabetes-Cancer	251
Heart-Diabetes-Cancer-Hypertension	749
Heart-Diabetes-Hypertension	506
Heart-Diabetes-Stroke	**19**
Heart-Diabetes-Stroke-Cancer	**32**
Heart-Diabetes-Stroke-Cancer-Hypertension	343
Heart-Diabetes-Stroke-Hypertension	111
Heart-Hypertension	1787
Heart-Stroke	**89**
Heart-Stroke-Cancer	154
Heart-Stroke-Cancer-Hypertension	360
Heart-Stroke-Hypertension	197
Hypertension	8475
Stroke	166
Stroke-Cancer	199
Stroke-Cancer-Hypertension	376
Stroke-Hypertension	284

Obesity status was based on the values of the body mass index (BMI), which is the body weight in kilograms divided by the square of the body height in meters. We used the standard World Health Organization classification system to categories individuals as Underweight (BMI < 18.5), Normal ($18.5 \leq$ BMI < 25), Overweight($25 \leq$ BMI < 30) and Obese (BMI $\geq$ 30).

Additional covariates used in the analysis are the five chronic conditions at baseline, age category, gender, income, work status, private health insurance status, Body Mass Index (BMI) and smoking status.

The SEEF study includes many more variables (such as education, dietary habits or family history) that could be used in the analysis but we have restricted ourselves to this set because we found that adding more variables did not significantly improve the predictions.

Since individuals were recruited in the 45 and Up Study over a period of few years the interval between interviews is not always the same, resulting in follow-up data being collected between 2 and 4 years after baseline. Therefore we also included as a covariate the time to follow up, which on average was 2 and half years. The summary statistics for the covariates used in the model are shown in Table 2.

Table 2. Summary statistics of the SEEF Study. All quantities measured at baseline except when reported otherwise. Quantities in parenthesis are proportions.

Age	[45,50]	(50,55]	(55,60]	(60,65]	(65,70]	(70,75]	(75,80]	(80,85]	(85,100]
	8902	10302	10196	9115	8007	5197	4132	3062	926
	(0.15)	(0.17)	(0.17)	(0.15)	(0.13)	(0.09)	(0.08)	(0.05)	(0.01)

Income	>20K	20K-30K	30K-40K	40K-50K	50K-70K	70K+
	13954	7874	6510	5769	8208	17524
	(0.23)	(0.13)	(0.10)	(0.10)	(0.14)	(0.30)

Health Insurance	No Private	Veteran	Health CareCard	Private Extras	Private No Extras
	9236	1045	9186	30970	9402
	(0.15)	(0.02)	(0.15)	(0.52)	(0.16)

BMI	Normal	Obese	Overweight	Underweight
	22780	12344	23950	765
	(0.38)	(0.21)	(0.40)	(0.01)

Smoking	Not smoker	Quit smoking	Smoker
	34910	21485	3444
	(0.59)	(0.35)	(0.06)

Work status	Full	Not working	Part time
	18002	29608	12229
	(0.30)	(0.50)	(0.20)

Gender	Female	Male
	32128	27711
	(0.54)	(0.46)

Conditions	Heart	Diabetes	Stroke	Cancer	Hypertension
	7059	4441	1256	22283	20598
	(0.12)	(0.07)	(0.02)	(0.37)	(0.34)

Conditions at follow-up	Heart	Diabetes	Stroke	Cancer	Hypertension
	10564	5973	2036	28278	26233
	(0.18)	(0.10)	(0.03)	(0.47)	(0.44)

3 Methodology

3.1 Multivariate Probit

Let us denote by $Y_{i\alpha}^{(1)}$ the binary variable indicating the presence at follow-up of chronic condition α for individual i, where $i = 1, 2, \ldots, N$ with $N = 60,000$, and $\alpha = \{$heart disease, diabetes, hypertension, stroke, cancer$\}$. Let us also denote by

$Y_{i\alpha}^{(0)}$ the corresponding variable measured at baseline, and by $\mathbf{Z}_i \in R^d$ a vector of other covariates measured at baseline. To simplify the notation we denote by $\mathbf{Y}_i^{(1)}$ ($\mathbf{Y}_i^{(0)}$) the vectors whose components are $Y_{i\alpha}^{(1)}$ ($Y_{i\alpha}^{(0)}$).

The MVP model is a latent variable model with the following specification:

$$\hat{\mathbf{Y}}_i^{(1)} = \Gamma \mathbf{Y}_i^{(0)} + \Theta \mathbf{Z}_i + \epsilon_i, \qquad Y_{i\alpha}^{(1)} = 1 \text{ if } \hat{Y}_{i\alpha}^{(1)} > 0, 0 \text{ otherwise} \qquad (1)$$
$$\epsilon_i \sim \mathcal{N}(0, \Sigma)$$

where Γ and Θ are matrices of coefficients, of dimensions 5×5 and $5 \times d$ respectively, that need to be estimated. The key to the MVP model of Eq. 1 is the presence of the 5×5 (unknown) covariance matrix Σ. The off-diagonal elements of its inverse capture the correlations across chronic conditions and the fact that developing, say, heart disease and diabetes are not independent events. Prediction of the MVP model are performed probabilistically, by feeding samples of the multivariate normal distribution $\mathcal{N}(0, \Sigma)$, one for each individual, in Eq. 1.

The estimation of the full MVP model is notoriously computationally intensive, although recent advances in computational methods [6] make it much more approachable. For the purpose of our experiments we have developed an approximation of the traditional method in which we use observed correlation among chronic conditions to approximate the matrix Σ, which makes the estimation of the model much simpler. Since we have not observed deterioration in performance by using the approximate method, all the experiments performed for the production of this paper have been performed using the approximation rather than the full implementation.

3.2   Support Vector Machines

Support Vector Machines (SVMs) have been around the machine learning community for more than 20 years now [9], and for the sake of brevity we simply refer the reader to standard textbooks and references [10,11]. SVMs have many attractive features, but one that should be emphasized in the context of this paper is that, unlike MVP, they do not rely on distributional assumption regarding the process that generates the data. Instead, SVMs relies on two key modeling choices: one is

1. the parameter (usually denoted by C) that controls the penalty associated with the misclassification of a data point;
2. the kernel, that is associated with the choice of the (possibly infinite dimensional) feature space onto which the input variables are projected [12]. For the purpose of this paper we have mainly experimented with polynomial kernels of the form $K(\mathbf{x}_i, \mathbf{x}_j) = (1 + \mathbf{x}_i \cdot \mathbf{x}_j)^p$, that are uniquely parametrized by the degree p.

SVMs were originally designed for binary classification problems, but several extensions exist that allow to deal with multi-class problems.

In the R package we use for the SVM implementation, Kernlab, there are several options for dealing with multi-class problems [13,14]. We found that for this problem best results were obtained by using the "one vs one" approach, in which one trains $K(K-1)/2$ binary classifiers (with $K = 32$ in our case). Each of the classifier separates one class from another class, and in order to classify a new sample, all classifiers are applied and the class that gets the highest number of votes is selected. While it is not fully clear why the "one vs one" approach worked better than the alternatives (such as the "one vs all" approach [15]), the fact that in this particular application many of the events we are trying to predict are quite rare seems to play a role, since it can lead to very imbalanced data sets.

4   Experimental Results

4.1   Performance Evaluation Metrics

We used a 10-fold cross-validation approach to estimate the performance of the MVP and SVM methods. The full data sets was first randomly partitioned in 10 subsets of equal size (approximately 6,000 data points each). For each of the 10 replication trials we withhold one of the 10 partitions and use it for testing, while the remaining 9 partitions are used for training. For each of the 10 trials we compute 4 performance measures, and we report the average of the performance measures over the 10 replications.

As performance measures we report sensitivity and specificity, since they are the ones most commonly used in health studies, as well as accuracy and the F1 score. We report the definitions below, where TP, TN, FP and FN refer to the total number of true positives, true negatives, false positives and false negatives respectively.

$$\text{Sensitivity} = \frac{TP}{TP+FN} \tag{2}$$

Sensitivity (or true positive rate, or recall) is important because it measures the ability to identify who is going to develop the disease.

$$\text{Specificity} = \frac{TN}{TN+FP} \tag{3}$$

Specificity (or true negative rate) is important because it measures the ability to identify who is *not* going to develop the disease.

$$\text{Accuracy} = \frac{TP+TN}{TP+FP+TN+FN} \tag{4}$$

Accuracy indicates how many samples are correctly classified overall. Accuracy can be misleading when the dataset is imbalanced. Therefore an alternative performance measure is the F1 Score, defined as:

$$\text{F1 Score} = 2 \times \frac{pr}{p+r} \tag{5}$$

where p is the precision and r is the recall (or sensitivity). Here precision is defined as the ratio of true positives (TP) to all predicted positives ($TP + FP$). Since the F1 score is the harmonic mean of precision and recall a high score is obtained when precision and recall are both high.

4.2   Results

The average of the performance measures over the 10 replication sets for Multi-class SVMs (MSVMs) and MVP are shown in Figs. 1 and 2. In Fig. 1 we report specificity and sensitivity for both methods. The key message of this figure is that while the specificity of the two methods are comparable, the sensitivity of MSVM is, on average about 12 % points better than the one of multivariate probit. Since sensitivities are in general not very high, this translates in a large relative improvement, of approximately 30 %.

A similar pattern is seen on accuracy and F1 scores. With very few exceptions SVMs are more accurate than MVPs, although by not too much. That the difference is not great relates to the fact that in most cases the classification

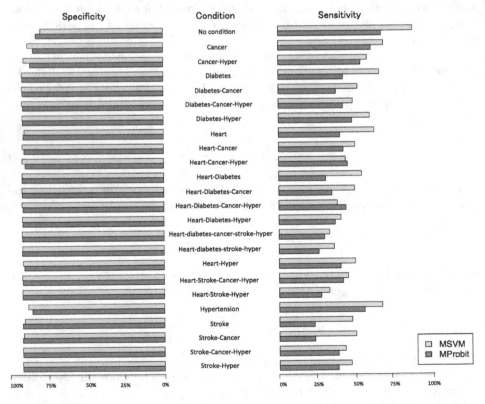

Fig. 1. Comparison between MSVM and MVP using 10-fold cross-validation: sensitivity and specificity.

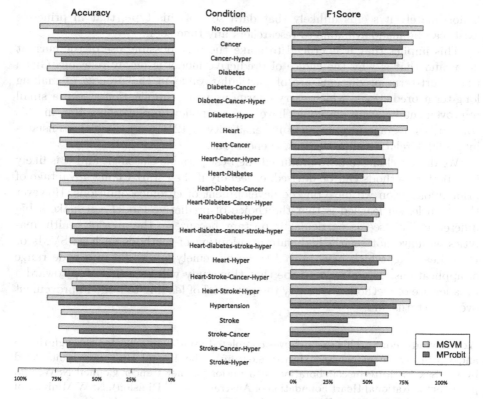

Fig. 2. Comparison between MSVM and MVP using 10-fold cross-validation: accuracy and F1 score.

problem is quite imbalanced, for which accuracy is not a good performance measure. The F1 score shows show larger differences between SVMs and MVPs, which is not surprising since a component of the F1 score is the sensitivity of the method, that is greatly improved using MSVMs.

5   Lessons Learned

Few lessons have emerged from this study. First of all, independently of which method we use, predicting who is going to develop some combination of chronic conditions in the near future, based on a handful of individual characteristics and the current chronic conditions, is quite hard. While maintaining specificity rates above 90%, most of the sensitivity rates, obtained using MSVMs, fell within 50% and 75%.

In our experience including additional risk factors, such as diet or family history, will only lead to marginal improvements. What is likely to have a major impact on the predictive ability of any method is a more accurate measurement of people's health status, such as actual results of pathology and imaging tests.

Unfortunately it seems unlikely that data sets of this type, that in principle exist, can be made available to researchers any time soon.

This implies that it is crucial to make the best possible use of the current data, and that is why the choice of predictive model is highly relevant. Given that short-term predictions are of particular value in the process of making long-term predictions, which carry enormous policy implications, even a small improvement in accuracy could have serious policy implications. Put in this context, an average improvement in sensitivity of 12 % points, which translates into a 30 % relative improvement, is enormous.

We do not claim to have produced the best possible classifier, and it is likely that better methods can be devised, especially if they start taking advantage of prior information we have on the development of chronic conditions. However the main lesson learned is that the choice of predictive model can make a big difference. This seems particular important because in the area of health analytics we have not seen a high rate of adoption of methods such as SVMs or Deep Learning, which have proved to be extremely successful in a wide range of applications. Therefore we hope that this study will be a first step toward a broader use of methods that carry the potential of leading to large improvement over the status quo.

Acknowledgment. This research was completed using data collected through the 45 and Up Study (http://www.saxinstitute.org.au). The 45 and Up Study is managed by the Sax Institute in collaboration with major partner Cancer Council NSW, and partners the National Heart Foundation of Australia (NSW Division), NSW Ministry of Health, beyondblue, NSW Government Family & Community Services Carers, Ageing and Disability Inclusion, and the Australian Red Cross Blood Service. We thank the many thousands of people participating in the 45 and Up Study. We also thank Capital Markets CRC,that has sponsored this research.

References

1. World Health Organization: Preventing Chronic Diseases. A Vital Investment, World Health Organization, Geneva (2005)
2. Goldman, D., Zheng, Y., Girosi, F., Michaud, P.-C., Olshansky, J., Cutler, D., Rowe, J.: The benefits of risk factor prevention in americans aged 51 years and older. Am. J. Public Health **99**(11), 2096–2101 (2009)
3. Lymer, S., Brown, L., Duncan, A.: Modelling the health system in an ageing Australia, using a dynamic microsimulation model. Working Paper 11/09, NATSEM at the University of Canberra, June 2011
4. Boyle, J.P., Thompson, T.J., Gregg, E.W., Barker, L.E., Williamson, D.F.: Projection of the year 2050 burden of diabetes in the US adult population: dynamic modeling of incidence, mortality, and prediabetes prevalence. Popul. Health Metr. **8**(1), 29 (2010)
5. Greene, W.: Econometric Analysis. Pearson Education, New Jersey (2003)
6. Cappellari, L., Jenkins, S.: Calculation of multivariate normal probabilities by simulation, with applications to maximum simulated likelihood estimation. Stata J. **6**(2), 156–189 (2006)

7. Collaborators, U.S.: Cohort profile: the 45 and up study. Int. J. Epidemiol. **37**(5), 941 (2008)
8. Mealing, N., Banks, E., Jorm, L., Steel, D., Clements, M., Rogers, K.: Investigation of relative risk estimates from studies of the same population with contrasting response rates and designs. BMC Med. Res. Methodol. **10**(1), 26 (2010)
9. Cortes, C., Vapnik, V.: Support-vector networks. Mach. Learn. **20**(3), 273–297 (1995)
10. Scholkopf, B., Smola, A.J.: Support Vector Machines, Regularization, Optimization, and Beyond. MIT Press, Cambridge (2001)
11. Cristianini, N., Shawe-Taylor, J.: An Introduction to Support Vector Machines and Other Kernel-based Learning Methods. Cambridge University Press, Cambridge (2000)
12. Smits, G.F., Jordaan, E.M.: Improved SVM regression using mixtures of kernels. In: Proceedings of the 2002 International Joint Conference on Neural Networks, IJCNN 2002, vol. 3, pp. 2785–2790. IEEE (2002)
13. Crammer, K., Singer, Y.: On the algorithmic implementation of multiclass kernel-based vector machines. J. Mach. Learn. Res. **2**, 265–292 (2002)
14. Weston, J., Watkins, C.: Support vector machines for multi-class pattern recognition. ESANN **99**, 219–224 (1999)
15. Rifkin, R., Klautau, A.: In defense of one-vs-all classification. J. Mach. Learn. Res. **5**, 101–141 (2004)

Method of Learning Malware Behavior Scripts by Sequential Pattern Mining

A.V. Moldavskaya[✉], V.M. Ruvinskaya, and E.L. Berkovich

Department of System Software, Odessa National Polytechnic University,
1, Shevchenko Avenue, Odessa 65044, Ukraine
amme4od@mail.ru, iolnlen@te.net.ua, evg.berkovich@gmail.com

Abstract. Scripts are the knowledge representation model. To our knowledge, there were no machine learning methods for it. In this work we propose a method of discovering script goals and putting them into an order. It is based on sequential pattern mining and regular expressions. This method has been validated by experiments set on malware behavior data. The results show that the discovered goals and their order correspond with expected malware behavior.

Keywords: Scripts · Knowledge representation · Sequential pattern mining · Malware

1 Introduction

Scripts, as the form of knowledge representation, were first proposed during 1970s [1], and yet since that time there wasn't much progress in projecting automated systems based on scripts. The reason for this is in the lack of methods of machine learning for this form of knowledge representation. This leaded to the limited usage of the whole scripts knowledge model. Our aim is to develop such a method. As far as we know, the literature has not discussed the usage of machine learning methods for obtaining a scripts knowledge model.

In Sect. 2, we explore the relevant work which started the theory of scripts. In Sect. 3, we analyze the requirements for a script-based knowledge system. In Sect. 4, we give the description of our method which suggests using sequential pattern mining for learning scripts from a sequence set. As the example field of knowledge, we choose the malicious software and its behavior. The behavior of malicious software often follows the so-called lifecycle, as found, for example, in [13]. The lifecycle describes the typical action sequence performed by a malware of a certain class. This resembles the scripts, therefore we find it appropriate to use scripts for modeling the malware lifecycle. Currently, the knowledge about generalized malware lifecycles (to not be confused with precise behavior knowledge) is only processed manually, as we show in our previous work [2]. Therefore, our choice of knowledge field is, firstly, for explaining and then experimentally testing our method; and secondly, for solving the issue with automated extraction of malware lifecycle models from raw data. In Sects. 4.1, 4.2 and 4.3 we

© Springer International Publishing Switzerland 2016
A. Gammerman et al. (Eds.): COPA 2016, LNAI 9653, pp. 196–207, 2016.
DOI: 10.1007/978-3-319-33395-3_14

provide and explain the results of experiments, showing sequential pattern mining of malware behavior and script learning based on the sequential patterns. Finally, in Sect. 5 we conclude the paper.

2 Related Work

The idea of scripts has two different interpretations: by Schank and by Minsky. Both of them we're comparing in [2]. Basically, Minsky's concept is closer to the frame knowledge model while Schank's is more free-standing. In this paper, we will follow Schank's model. Firstly, we make a short summary of this model, as described in [1–3].

Script is the structured knowledge representation model, used for representing the typical behaviors in some context. The structure of script resembles the one of frames model, except its purpose is to describe the sequences of actions or events. Scripts are used for modeling the human perception, human inference and natural language comprehension. They allow considering the context of the events, which opens the possibility to fill in for the missing parts of incoming information.

Scripts consist of a wide range of component elements, with the most fundamental one being the goal. By achieving goals, the described behavior moves on. The goal can include following elements [1]:

1. The action, which is necessary to complete for achieving the goal
2. The object affected by the action
3. The actor, or the source of performing the action
4. The direction of the action

A basic script is the sequence of goals performed by one actor, representing the single typical behavior in a pre-described context. The generalized goals can be broken into more specific subgoals. A more complicated script can represent multiple behaviors, thus, can include multiple actors. In this case, the notion of role is introduced. The role is the typical behavior of the actor, performed in a certain context. It must be noted that the name of the actor is not important for a script-based analyzing system, and only the role is analyzed.

By further complicating the script, it's possible to group the large, branching goal sequences into scenes. A scene is a group of goals, united by being related to a certain area of the script's context. Scenes describe the typical behaviors for such areas.

Schank proposes enriching the script model even further, by:

1. Adding parameters (modifiers, etc.)
2. Creating categories of scenes and roles
3. Establishing the relations between roles
4. Adding obstacles and distractions

Obstacles and distractions are two wide class of elements found in the incoming information. They are not included in the scripts, because they represent an unexpected external action. Obstacles are the elements which forbid the achievement of the current goal. Distractions are the new unknown goals which interrupt the script following process.

The described features show that scripts open wide possibilities for hierarchic organization of its elements, while still maintaining their order. Following this overview, we compared scripts with alternative existing knowledge representation models, as described in [5]. Frames model is the closest to scripts. The difference is, the objective of frames is to deeply describe the context, while the objective of scripts is to deeply describe the sequences of actions taken within the context. Production rules resemble scripts as well, in the sense that both represent the sequences of some kinds of events. Yet scripts, as it was shown above, feature a number of context-describing parameters, which do not exist in the production rules model. As the result, the inference based on production rules is unable to consider the context – which is exactly the purpose of script-based inference [1]. Semantic networks do not quite resemble scripts, as they do not represent the sequences, just the semantic connections. The common part between scripts and semantic networks is the support for hierarchy.

We've compared the scripts with other knowledge representation models following the comparison methodic described in [6]. As the result, we've marked out the merits and demerits of representing knowledge with scripts. Among the merits, the clearness for human perception and the universality for knowledge fields can be noted. Among the dismerits, the most significant one is the complexity of model learning and the lack of methods to perform it. The main problem of scripts, as Schenk mentions it [3], is the necessity to manually input large amount of data to create the script set. Nevertheless, the manually made scripts are used for building models in the fields related to human perception, linguistics mostly [7,8].

3   The Features of a Script-Based Expert System

The purpose behind a script-based expert system, reasoning system or a script-based decision making system is to expand the incomplete input data by inference, thus discovering new knowledge. For that, the system must solve the following tasks:

1. Accept the set of learning samples, convert it into a set of scripts and store the resulting set
2. Estimate the context of the input and choose the appropriate script, on which the inference will be based
3. Make correlations between the input an the chosen script, while completing the lacking information via inference
4. Take obstacles and distractions into account, in case the area of application demands it

The most important function of such a system is its ability to perform machine learning, meaning that it would build scripts from raw data.

4   The Machine Learning Method for Scripts

Humans are able to interpret the context, because they know the typical regularities and patterns. These regularities are known from observing the environment and the typical events, which occur regularly. This leads to the idea that a script, being the representation of typical events or actions, should be learned from a dataset of events or actions. The machine learning process should be able to derive the generalized, typical sequences. The dataset must have a number of action sequences which happened in the same context in question, or were performed by the role in question. All of these sequences can be different in details, but generally they, being typical, will represent the same situation.

Therefore, to create the method of script machine learning it is necessary to solve two tasks:

1. Discover the regularities, or patterns from the action sequence dataset and convert them into goals and subgoals.
2. Arrange the goals and subgoals in sequence corresponding to the original dataset to obtain a script.

In this paper, the task of finding the regularities in sequences is solved with data mining methods – more specifically, sequential pattern mining methods. These methods did not exist in the years when scripts theory was proposed and expanded, and the computation capabilities of computer systems did not allow processing large data – which is not the question now. Firstly, we make an overview of sequential pattern mining and explore its potential for script learning.

4.1   Sequential Pattern Mining Overview

Sequential pattern mining is the branch of data mining. The object of analysis for sequential pattern mining is a set of sequences. A sequence is an ordered list of itemsets. An itemset is a non-empty set of items [9]. An example of a sequence: $S = \langle \{a\}, \{a,b,c\}, \{b\}, \{b, c\}, \{a, d\} \rangle$. If an itemset is composed of items representing the events, then the events within the itemset would be the ones occurring simultaneously. A sequence representing events only coming one after another will be represented with itemsets of one item each.

The goal of sequential pattern mining is to obtain common subsequences from the initial sequence set [10]. These subsequences are named the sequential patterns. The evaluation of how common a subsequence is depends on the chosen method of mining, but generally it is based on the support parameter $sup(P)$, where P stands for the supported pattern. The support indicates how many sequences from the initial set contain a candidate subsequence, i.e. the subsequence which is checked for being a sequential pattern. Once the support for a subsequence in question hits a certain pre-defined ceiling, the subsequence is considered to be a sequential pattern, and gets placed into the output set.

The main problem of sequential pattern mining, as [10] summarizes, is the output set size. The large amount of patterns makes it difficult to interpret the results. Because of that, it's common to only seek for the patterns with more narrow definitions, which also tightens the requirements for a candidate subsequence to qualify as a sequential pattern. For example, the sequential pattern P_a is closed, if there is no other sequential pattern P_b which would be a supersequence for P_a while $\sup(P_a) = \sup(P_b)$ [11]. Narrowing the definition even further, we can get the maximal sequential pattern, as described in [10, 11]. The sequential pattern Pa is maximal, if there is no other sequential pattern P_b which would be the supersequence for P_a. The support is not considered while mining the maximal sequential patterns. So, from a candidate sequences set $\langle P_a, P_b \rangle$, if $\sup(P_a) > \sup(P_b)$, and $P_a \subset P_b$, then P_a and P_b are both closed sequential patterns, but only P_b is a maximal sequential pattern.

4.2   Method of Learning Scripts from Sequential Patterns

The purpose of the following method is to discover and build a set of hierarchic scripts consisting of goals and subgoals in a pre-described context. The method solves both tasks that we've set in Sect. 3. The described method is explained on malware behavior data, although can be extended to handle any kind of behavior data presented in sequences of actions.

The idea behind the method is to apply a two-steps approach. In Step 1, the generalized goals in the form of patterns are discovered by applying sequential pattern mining to the behavior set. The behavior set consists of action sequences. In Step 2, these goals are put in order by referring back to the order of similar actions in the behavior set. The result of these two steps is a set of two-layered hierarchic scripts. The steps can be reiterated for adding new layers to the hierarchy by discovering even more generalized goals.

Let input set D_b^0 of size m be a 0th layer behavior set. In our example, it is, specifically, a malware behavior reports set. Every report is a sequence of WinAPI function names, called by a malware during its execution – i.e. the actions. D_b^0 is analyzed with a chosen algorithm of sequential pattern mining, and the pattern set D_p^1 of size n is discovered as the result. The exact value of n depends on the chosen algorithm and the initial parameters of mining process. We will call $D_p^1 = \{p_1^1 \ldots p_n^1\}$ a 1st layer patterns set, its items we will call 1st layer patterns. The patterns consist of items, specifically WinAPI function names in our case. These items we will call 0th layer items. Both layers are shown on Fig. 1. In terms of scripts, we consider D_p^1 being a set of generalized goals consisting of actions from the 0th layer.

In step two, we propose using regular expressions to solve the task of discovering the order of goals. For the discovered set $D_p^1 = p_1^1 \ldots p_n^1$, a regular expression is created:

$$p_1^1 | p_2^1 | \ldots | p_n^1 \text{ (R1)}$$

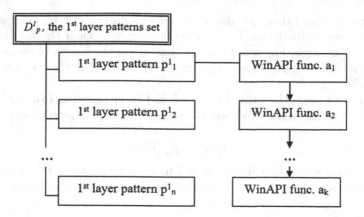

Fig. 1. An example of $1st$ level pattern set, with one of the patterns of length k being expanded on $0th$ layer

R1 is applied to the behavior set D_b^0. This allows discovering the behavior set D_b^1 of size m_1, consisting of ordered sequences of sequential patterns as items. In terms of scripts, D_b^1 consists of scripts, which have a number of goals put in order. Every one of these goals is achieved by performing a sequence of subgoals, each of a size of one action. Therefore, D_b^1 is the output of our method and the set of scripts.

For creating the next layer of the script, D_b^1 is taken as the initial behavior set. After applying a chosen sequential pattern mining algorithm, the set of patterns D_p^2 of size n_2 is received: $D_p^2 = \{p_1^2 \dots p_n^2\}$. D_p^2 consists of sequences made from 1st layer patterns as the items. These sequences, $\{p_1^2 \dots p_{n2}^2\}$, we will call 2nd layer patterns (see Fig. 2).

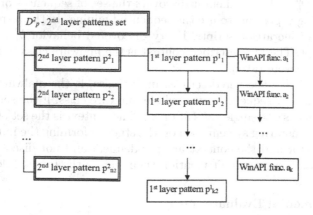

Fig. 2. An example of $2nd$ layer pattern set, with one of the $2nd$ layer patterns of length $k2$ being expanded on $1st$ layer. One of this pattern's items, being a pattern itself, is expanded into $0th$ layer.

It is to be noted that the algorithm for sequential pattern mining on this step is not necessarily the same as on the previous one, when D_p^1 was discovered. The work of an algorithm and its results depend on the length of sequences and the input set size. For D_b^2, as we experimentally discovered, these will be smaller than for D_b^1.

At the next step, the proper order of 2nd layer patterns from D_p^2 is to be discovered. For the discovered set $D_p^2 = \{p_1^2 \ldots p_{n2}^2\}$, a regular expression is created:

$$p_1^2 | p_2^2 | \ldots | p_n^2 \ (\text{R2})$$

The regular expression R2 is applied to the behavior set D_b^1. The result is the behavior set D_b^2 of size m_2. This set consists of sequences made from 2nd layer patterns as items (see Fig. 3.):

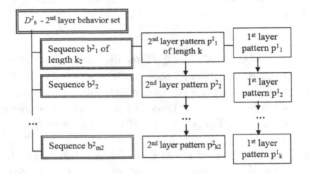

Fig. 3. Behavior set, expanded level by level

To sum it up, the method uses the sequential patterns set D_p^1 discovered from sequences set D_b^0 and from them it discovers the set of sequences of higher level D_b^1, thus making it possible to use the sequential mining. In the terms of scripts, D_b^2 is the set of hierarchic scripts. Every discovered behavior $b_1^2 \ldots b_{m2}^2$ is an hierarchic script. The pre-described context is the malware class from which the behavior reports were taken.

The actor of the discovered scripts, in our case, is the malware which performed the behavior sequences (or a whole malware class, if the set is based on different specimens of a single class at once). The context is the set of parameters for the report-generating system used on the stage of forming the initial behavior set D_b^0. The actor and the context are pre-defined, and not discovered via the mining process, hence we don't mention them across the method description.

4.3 Experimental Evaluation

This subsection demonstrates the application of our method on real raw data on malware behavior. The experiments were done in two phases. Firstly, we discovered the sequential patterns from sequences representing malware behaviors.

For that, we use the existing sequential pattern mining methods. Secondly, we built the scripts based on these discovered patterns. This is done with the usage of our method, as proposed in Sect. 4.2.

The experiment was set with the usage for following data and instruments. The initial behavior datasets corresponded with malware classes. They were formed from the collection of malware behavior reports obtained in [12], which contained 3 157 reports in XML format, presented as WinAPI names sequences. The collection embraces the phases of malware lifecycle which do not require network interaction. The data went through simple preprocessing before applying the sequential pattern mining methods. For creating an input dataset D_b^0 for every malware class, we purged the reports of unknown class malwares as well as the reports of malware families consisting of less than 3 reports. As the result, the input datasets were of following size: Backdoor – 595, Virus – 94, Worm – 224, P2P-Worm – 179, Trojan – 277.

The chosen algorithms of sequential pattern mining were CloSpan and ClaSP, which both discover closed sequential patterns. The difference is the method of data mapping. CloSpan maps the data into a tree, while ClaSP utilizes vertical data representation, which allows higher speed but generates different candidate sequences [11].

We've set the minimal pattern length as 3, meaning that shorter patterns will be purged from the output set. The experiments were held independently from each other, meaning that the discovered patterns were not purged by overlapping. The quantities of 1st layer patterns discovered by CloSpan with according support settings are shown in Table 1.

Table 1. 1st layer patterns, discovered by CloSpan

Class	Total	Patterns discovered, support 100%–50%					
		100%	90%	80%	70%	60%	50%
Backdoor	595	0	0	1	49	389	N/A
Virus	94	0	0	0	2	9	12
Worm	224	0	1	8	11	90	998
P2P-Worm	179	0	3	45	253	688	N/A
Trojan	277	0	10	21	38	243	626

The quantities of 1st layer patterns discovered by ClaSP with according support settings are shown in Table 2.

Next, we set an experiment to discover the 2nd layer patterns. For the input, we used the sets of patterns discovered by ClaSP from P2P-Worm and Trojan malware classes, support value 50%. These were the largest sets obtained at the previous experiment. We applied ClaSP and CloSpan with support 100%-30%. The quantities of discovered 2nd layer patterns as the result of this experiment are shown in Tables 3 and 4.

Table 2. 1st layer patterns, discovered by ClaSP

Class	Total	Patterns discovered, support 100%-50%					
		100%	90%	80%	70%	60%	50%
Backdoor	595	0	0	23	49	389	N/A
Virus	94	0	3	3	5	14	26
Worm	224	0	0	0	11	90	998
P2P-Worm	179	0	3	45	253	688	1268
Trojan	277	0	10	31	71	318	946

Table 3. The results of 2nd layer pattern mining for P2P-Worm class

Algorithm	2nd layer patterns discovered							
	100%	90%	80%	70%	60%	50%	40%	30%
ClaSP	0	0	0	0	0	0	1	2
CloSpan	0	0	0	0	0	1	1	2

Table 4. The results of 2nd layer pattern mining for Trojan class

Algorithm	2nd layer patterns discovered							
	100%	90%	80%	70%	60%	50%	40%	30%
ClaSP	0	0	0	0	0	1	1	3
CloSpan	0	0	0	0	0	1	1	4

Consider the specifical example of discovered patterns:

GetProcAddress() – InitializeSecurityDescriptor() – SetSecurityDescriptor-Dacl() – FreeSid() – GetProcAddress()

This pattern, being a WinAPI names sequence, demonstrates malware's way of working with a process security descriptor.

By further applying our method, as described in Sect. 3, we created scripts that had the discovered patterns as their goals and WinAPI names as subgoals. As an example, one of the scripts for Agent family of Backdoor class appeared as following:

1. GetACP – GetProcAddress – LoadLibraryA This pattern corresponds with the goal of receiving the operational system's code page.
2. GetProcAddress – InitializeAcl – AddAccessAllowedAce – InitializeSecurityDescriptor – RegCreateKeyExA – GetProcAddress – LoadLibraryA This pattern demonstrates how a backdoor achieves the goal of creating the access restriction, by using the security descriptor and putting it at a register key.
3. GetProcAddress – AllocateAndInitializeSid – InitializeAcl – AddAccessAllowedAce – InitializeSecurityDescriptor – RegCreateKeyExA – GetProcAddress – LoadLibraryA This pattern demonstrates how a backdoor achieves the

goal of adding an access restriction, by adding it into an access list, creating a security descriptor and putting it at a new register key.

4. GetProcAddress – AllocateAndInitializeSid – InitializeAcl – AddAccessAllowedAce – InitializeSecurityDescriptor – RegCreateKeyExA – FreeSid – GetProcAddress – LoadLibraryA

This pattern demonstrates achieving the same goal, but the security descriptor is also set free.

Therefore, if we consider the implications behind the discovered goals, we receive the following script (Fig. 4):

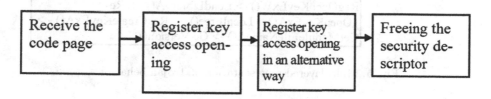

Fig. 4. The script for malwares of Agent family, Backdoor class.

Consider an example of a 2nd layer pattern discovered from Trojan class by ClaSP. It consists of two 1st layer patterns:

1. Pattern p_1^1: RegOpenKeyExW() – LoadLibraryA() – RegOpenKeyExA() – Local-Free() – RegCreateKeyExA() – GetSystemMetrics() – GetModuleFileNameA()
2. Pattern p_2^1 : RegOpenKeyExW() – LoadLibraryA() – RegOpenKeyExA() – Local-Free() – RegCreateKeyExA() – GetModuleFileNameA() – GetVersion()

This 2nd layer pattern demonstrates how Trojan malwares incor-porate themselves into an operational system by consecutively trying to edit the register in two different ways, as shown in Fig. 5:

Another 2nd layer pattern, consisting of 3 1st level patterns, shows the similar behavior.

1. RegOpenKeyExW() – LoadLibraryA() – RegOpenKeyExA() – LocalFree() – Reg-CreateKeyExA() – LoadLibraryA() – RegCloseKey()
2. RegOpenKeyExW() – LoadLibraryA() – RegOpenKeyExA() – LocalFree() – Reg-CreateKeyExA() – RegOpenKeyExW() – LoadLibraryA()
3. RegOpenKeyExW() – LoadLibraryA() – RegOpenKeyExA() – LocalFree() – Reg-CreateKeyExA() – LoadLibraryA() – RegOpenKeyExA()

Therefore, these discovered patterns show different chains of complex actions by which Trojan malwares try to achieve their general goal. The discovered scripts can be used for better understanding of malware behavior and for intellectually malware detection systems.

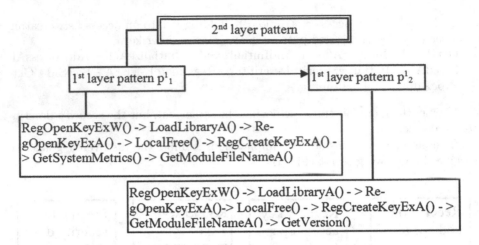

Fig. 5. Multi-layered representation of Trojan behavior

5   Conclusion

In this paper, we discussed the method of applying machine learning to Schank's scripts model. The previous studies of Schank and co-authors [1, 3, 4] describe in detail the main concepts and elements for a script-based knowledge model, yet they provide no methods of machine learning for it. By reviewing the practical applications of scripts (for example, [7, 8]), we discovered that scripts are still being formed manually. Our goal was to create a machine learning method which could've solve the task of forming the most basic and necessary script elements from raw data sequences. For that, we first compared the scripts with other knowledge representation models. Second, we formulated the requirements for a self-learning script-based computer system. Basing on these requirements, we propose the method of script goal mining and ordering. That allowed creating the hierarchic scripts automatically.

The experiments, set on malware behavior dataset, show the following results. 1st layer patterns obtained, for 5 malware classes and support 90 %: 1, 3 10 (3 classes); support 80 %: 1 to 45 (4 classes); support 70 %: 2 to 253 (5 classes); support 60 %: 9 to 68 (5 classes); support 50 %: 12 to 1268. 2nd layer patterns obtained, for 2 malware classes, with support 50 %: 1 each; support 40 % - 1 each, support 30 %: 2 to 4 for each class. With the proposed method of script learning, the discovered patterns were united into scripts. By analyzing the specific examples, we saw that the discovered patterns and resulting scripts make sense and correspond with real malware actions.

The described method was tested only on malware behavior data, however, it can be of interest in analyzing and mining the behavior-related data of any nature. There is a wide range of possible future research based on the suggested method. Our next steps will be towards building a script-based, malware behavior detecting expert system; as well as developing an approach for discovering cyclic script goals among the normal ones. This research is currently in work, and seems to be leading us into developing a new kind of sequential patterns.

References

1. Schank, R.C., Abelson, R.P.: Scripts, plans and goals. In: Proceedings of the 4th International Joint Conference on Artificial Intelligence. IJCAI 1975, vol. 1 (1975)
2. Ruvinskaya, V.M., Moldavskaya A.V., Kholovchuk A.O.: Scenarii kak forma pred-stavlenia znaniy pro povedenie vredonosnyh program (Scripts as a form of representation of knowledge of malware behavior). Transactions of Kremenchuk Mykhailo Ostrohradskyi National University **4** (2014)
3. Schank, R.C., Riesbeck, C.K.: Inside Computer Understanding: Five Programs Plus Miniatures. Psychology Press, New York (2013)
4. Schank, R.C., Abelson, R.P.: Scripts, plans, and knowledge, pp. 151–157. Yale University (1975)
5. Joseph, Giarratano, Gary, Riley: Expert systems principles and programming, 2nd edn., p. 321. PWS Publishing Company, Boston (1998)
6. Krisilov, V.A., Poberezhnik S. M., Tarasenko R.A.: Sravnitelniy analiz mode-ley pred-stableniya znaniy v intellektual'nyh sistemah (Comparative analysis of the knowledge re-presentation models in artificial intelligence systems). Odes'kyi Politechnichnyi Universytet. Pratsi, vol. 2, p. 6 (2011)
7. Polatovskaya, O.S.: Freim-scenariy kak tip konceptov (Scene-frame as a concept type). ISTU Bulletin (2013)
8. Kushneruk, S.L.: Teoriya tekstovyh mirov: perspectivy issledovaniya reklamnoy kommunikacii. Political Linguistics **25** (2008)
9. Agrawal, R., Srikant, R.: Mining sequential patterns. In: 1995 Proceedings of the Eleventh International Conference on Data Engineering. IEEE (1995)
10. Gupta, M., Han, J.: Approaches for pattern discovery using sequential data mining. Pattern Discovery Using Sequence Data Mining: Applications and Studies, pp. 137–154 (2012)
11. Mabroukeh, N.R., Ezeife, C.I.: A taxonomy of sequential pattern mining algorithms. ACM Comput. Surv. (CSUR) **43**(1), 3 (2011)
12. Sami, A., et al.: Malware detection based on mining API calls. In: Proceedings of the 2010 ACM Symposium on Applied Computing. ACM (2010)
13. Chen, Z., et al.: Malware characteristics, threats on the internet ecosystem. J. Syst. Softw. **85**(7), 1650–1672 (2011)

Extended Regression on Manifolds Estimation

Alexander Kuleshov[1] and Alexander Bernstein[2,3(✉)]

[1] Skolkovo Institute of Science and Technology, Moscow, Russia
A.Kuleshov@skoltech.ru
[2] Institute for Systems Analysis, FRC CSC RAS, Moscow, Russia
a.bernstein@mail.ru
[3] Kharkevich Institute for Information Transmission Problems RAS,
Moscow, Russia

Abstract. Let f(X) be unknown smooth function which maps p-dimensional manifold-valued inputs X, whose values lie on unknown Input manifold **M** of lower dimensionality q < p embedded in an ambient high-dimensional space R^p, to m-dimensional outputs. Regression on manifold problem is to estimate a triple (f(X), $J_f(X)$, **M**), which includes Jacobian J_f of the mapping f, from given sample consisting of 'input-output' pairs. If some mapping h transforms Input manifold **M** to q-dimensional Feature space Y_h = h(**M**) and satisfies certain conditions, initial estimating problem can be reduced to Regression on feature space problem consisting in estimating of triple ($g_f(y)$, $J_{g,f}(y)$, Y_h) in which unknown function $g_f(y)$ depends on low-dimensional features y = h(X) and satisfies the condition $g_f(h(X)) \approx f(X)$, and $J_{g,f}$ is its Jacobian. The paper considers such Extended problem and presents geometrically motivated method for estimating both triples from given sample.

Keywords: Regression on manifolds · Regression on feature space · Input manifold estimation · Jacobian estimation · Tangent bundle manifold learning

1 Introduction

1.1 Regression Estimation

Regression estimation is a part of Statistical Learning whose general goal is finding a predictive function based on data [1–3]. Common regression estimation problem is as follows. Let T = f(X) be an unknown smooth mapping from its domain of definition **M** lying in input Euclidean space R^p to m-dimensional output Euclidean space R^m. Given training 'input-output' sample

$$\mathbf{Z}_n = \left\{ Z_i = \begin{pmatrix} X_i \\ T_i = f(X_i) \end{pmatrix}, i = 1, 2, \ldots, n \right\}, \tag{1}$$

the problem is to construct an estimator (learned function) f*(X) which maps the inputs X ∈ **M** to outputs f*(X) with small predictive error:

© Springer International Publishing Switzerland 2016
A. Gammerman et al. (Eds.): COPA 2016, LNAI 9653, pp. 208–228, 2016.
DOI: 10.1007/978-3-319-33395-3_15

$$f^*(X) \approx f(X), \tag{2}$$

and, thus, can be used to predict output $f(X)$ for new 'previously unseen' input $X \in M$.

There are various approaches and methods for reconstruction of an unknown function from the training sample, such as least squares techniques (linear and non-linear), artificial neural networks, kernel nonparametric regression and kriging, SVM-regression, Gaussian processes, Radial Basic Functions, Deep Learning Networks, Gradient Boosting Regression Trees, Random Forests, etc. [4–9].

In many applications, it is necessary to estimate not only function $f(X)$ but also its Jacobian matrix $J_f(X) = \nabla_X f(X)$. For example, many design tasks in Engineering are formulated as an optimization of given function $f_1(X)$ over design variable $X \in M$ under constraints defined in terms of other functions $f_2(X), f_3(X), \ldots, f_m(X)$, which together determine m-dimensional vector-function $f(X)$. Various gradient-based algorithms are used usually in solving of such optimization tasks. In general, the vector-function f is unknown and all available information about f is contained only in the sample Z_n (1). In this case, at first, the learned function $f^*(X)$, which in engineering applications is referred to as the surrogate function or meta-model, is constructed from the sample. Then, an initial optimization task is replaced by 'surrogate' optimization task about the constructed surrogate function $f^*(X)$ [10–12]. Therefore, for using gradient-based algorithms in this surrogate optimization task, we need either to provide proximity $J_{f^*}(X) \approx J_f(X)$ between $m \times p$ Jacobian matrices $J_f(X)$ and $J_{f^*}(X)$ of unknown and learned functions $f(X)$ and $f^*(X)$, respectively, or to construct sample-based $m \times p$ learned matrix $J^*(X)$ which accurately reconstructs the Jacobian $J_f(X)$:

$$J^*(X) \approx J_f(X) \tag{3}$$

for all points $X \in M$; then a pair $(f^*(X), J^*(X))$ is used in optimization procedures. Note that many regression methods don't include Jacobian estimation.

1.2 Regression on Manifolds Estimation

It is well known that if a dimensionality p of inputs X is large than many regression methods perform poorly due a statistical and computational 'curse of dimensionality': a collinearity or 'near-collinearity' of high-dimensional inputs cause difficulties when doing regression; reconstruction error in (2) cannot achieve a convergence rate faster than $n^{-s/(2s+p)}$ when nonparametric learned function estimates at least s times differentiable function $f(X)$ [13, 14].

Fortunately, in many applications, especially in imaging and medical ones, the high-dimensional inputs X occupy only a very small part $M \subset R^p$ in the high dimensional 'observation space' R^p whose intrinsic dimension q is small (usually, $q << p$).

Example Wing shape optimization is one of important problem in aircraft designing in which design variables include a number of p-dimensional detailed descriptions X of wing airfoils consisting of coordinates of points lying densely on the airfoils' contours [15]. In practical applications, the dimension p varies in the range from 50 to 200;

a specific value of p is selected depending on the required accuracy of airfoil description. But high-dimensional descriptions of 'real' aerodynamic airfoils occupy only a very small part of the 'airfoil-description' space R^p [16, 17] whose intrinsic dimension q varies in the range from 5 to 10.

The most popular model of high-dimensional data, which occupy a very small part of observation space R^p, is Manifold model in accordance with which the data lie on or near an unknown manifold (Data manifold) **M** of lower dimensionality q < p embedded in an ambient high-dimensional input space R^p (Manifold assumption [18] about high-dimensional data); typically, this assumption is satisfied for 'real-world' high-dimensional data obtained from 'natural' sources. In real examples, a manifold dimension q is usually unknown and can be estimated on the basis of given dataset randomly sampled from the Data manifold [19–22].

An estimation of unknown function f from the sample Z_n (1), whose domain of definition **M** = Supp(f) is unknown low-dimensional manifold (Input manifold, IM) embedded in high-dimensional input space R^p, is usually referred to as the Regression on manifolds estimation problem. We consider this problem in Manifold Learning framework in which various Data Analysis tasks are studied under the Manifold assumption about the processed data.

Note that term 'Manifold Learning' had previously been used usually in the 'narrow sense' and concerned only nonlinear Dimensionality Reduction tasks consisting in transforming of the manifold-valued data X into their low-dimensional representations (features) y = h(X) preserving certain chosen subject-driven data properties [23–25]; then the low-dimensional features of the original data can be handled efficiently, avoiding the curse of dimensionality. Thence, Manifold Learning is usually the first step in various Data Analysis tasks in which q-dimensional features are used in the reduced learning procedures instead of initial p-dimensional vectors [26].

Various Regression on manifolds methods in an explicit or an implicit form use Dimensionality reduction technique for discovering low-dimensional structure of the Input manifold. In [27, 28], kernel regression estimator is constructed directly on the manifold, using the true geodesic distance in both a determining the nearest neighbors and a constructing of kernel weights. Another approach [29] is to employ the usual Local Linear Regression technique in the ambient space R^p with regularization imposed on the coefficients in the directions perpendicular to the estimated tangent space to the Input manifold; the derivatives of the unknown function f were estimated in [30]. Manifold Learning tool called Manifold Adaptive Local Linear Estimator for the Regression [31] allows estimating the unknown function and its gradient; the tool explores Riemannian geometric structure of the Input manifold and constructs Local Linear Regression directly on estimated tangent spaces to the Input manifold, without knowing the geodesic distance and manifold structure. Bayesian nonparametric regression method for Regression on manifolds was proposed in [32, 33]; geodesic regression and polynomial regression on Riemannian manifolds are proposed in [34] and [35], respectively. A more general case concerning nonparametric regression between Riemannian manifolds (it means that output space R^m has dimension m > 1) is studied in [36] where minimization of regularized empirical risk is used for constructing of learned function.

In the paper, we consider Regression on manifold task as an estimating the triple $(f(X), J_f(X), M)$ with manifold-valued inputs $X \in M$ that includes an estimation of a support M of unknown function $f(X)$; covariant differentiation is used in $J_f(X)$.

1.3 Regression on Feature Space Estimation

As was written above, a final goal of Dimensionality reduction is a reducing of initial Data Analysis task concerning high-dimensional vectors $X \in R^p$ to similar 'reduced' task concerning low-dimensional features $y = h(X) \in R^q$ under given (or already estimated) dimension q of the IM M. In the considered Regression on manifold estimation task, after constructing the Feature space (FS) $Y_h = h(M) \subset R^q$, we consider 'reduced' task called Regression on Feature space estimation task about specific unknown function $g_f(y)$, which is defined on the FS Y_h and satisfies the proximity

$$g_f(h(X)) \approx f(X) \tag{4}$$

for all $X \in M$, that is follows: to estimate the function $g_f(y)$ from a Feature sample

$$Z_{n,h} = \{(y_{i,h} = h(X_i), g_f(y_{i,h}) = f(X_i)), i = 1, 2, \ldots, n\}. \tag{5}$$

If an estimator $g_f^*(y)$ for $g_f(y)$ defined on the FS Y_h satisfies proximity

$$g_f^*(y) \approx g_f(y) \tag{6}$$

for all $y \in Y_h$ then a function

$$f^*(X) = g_f^*(h(X)) \tag{7}$$

satisfies the required proximity (2) and, hence, gives a solution to the initial task.

Therefore, initial Regression on manifold estimation task about unknown function $f(X)$ with p-dimensional manifold-valued inputs X is replaced by reduced Regression on Feature space estimation task about unknown function $g_f(y)$ with q-dimensional inputs $y = h(X)$ and, thus, allows avoiding curse of dimensionality.

Because of this, we will consider Extended Regression on manifold task consisting in estimating of two triples $(f(X), J_f(X), M)$ and $(g_f(y), J_{g,f}(y), Y_h)$, in which $J_{g,f}(y)$ is $m \times q$ Jacobian matrix of the mapping $g_f(y)$; this estimation problem includes solving of Input manifold estimation problem and constructing of Embedding mapping h.

1.4 Input Manifold Estimation Problem

Before strict defining of the Input manifold estimation problem, consider above described optimization problem for unknown function $f(X)$ over high-dimensional manifold-valued inputs $X \in M$ (for example, wing shape optimization task [15]) which is reduced to surrogate optimization problem for learned function $g_f^*(y)$ over low-dimensional features $y \in Y_h$.

Let $y^* \in \mathbf{Y}_h$ be a solution of the surrogate optimization problem which does not belong in general case to the sample set $\mathbf{Y}_{n,h} = \{y_{i,h} = h(X_i), i = 1, 2, ..., n\}$; thus, we need to recover an 'optimal' value $X^* \in \mathbf{M}$ from its Out-of-Sample feature $y^* \notin \mathbf{Y}_{n,h}$ that satisfies the equation $h(X^*) = y^*$. Therefore, for further use of a solution obtained in Regression on Feature space estimation, we need constructing a Recovering mapping g from the FS $\mathbf{Y}_h$ to the input space R^p which meets the proximity

$$r(X) \equiv g(h(X)) \approx X \qquad (8)$$

between points $X \in \mathbf{M}$ and their recovered values $r(X)$ with small reconstruction error $\delta(X) = |r(X) - X|$ in the proximity (8); the mapping $r(X)$ from the IM $\mathbf{M}$ to the Input space R^p is a results of successively applying the embedding and recovering mappings to the inputs $X \in \mathbf{M}$.

The pair (h, g) determines q-dimensional Recovered Input manifold (RIM)

$$\mathbf{M}_{h,g} = r(\mathbf{M}) = \{X = g(y) \in R^p : y \in \mathbf{Y}_h \subset R^q\} \qquad (9)$$

covered by a single chart g and embedded in an ambient p-dimensional Input space R^p; the RIM $\mathbf{M}_{h,g}$ can be considered as an estimator of the IM $\mathbf{M}$ that meets proximity

$$\mathbf{M}_{h,g} \approx \mathbf{M} \qquad (10)$$

meaning small Hausdorff distance $d_H(\mathbf{M}_{h,g}, \mathbf{M}) \leq \sup_{X \in \mathbf{M}} \delta(X)$ between the manifolds.

The considered Input Manifold Estimation problem includes both a constructing the pair (h, g) satisfying the proximities (8), (10) and an estimating of $q \times p$ and $p \times q$ Jacobian matrices $J_h(X)$ and $J_g(y)$ of the mappings h and g, respectively; covariant differentiation is used in $J_h(X)$ when $X \in \mathbf{M}$.

1.5   Extended Regression on Manifolds Estimation: An Approach

If a solution (h, g) of the Input Manifold Estimation problem meets an additional 'functional' proximity

$$f(r(X)) \approx f(X), \qquad (11)$$

than we can replace initial Regression on Manifolds estimation problem about unknown function f(x) by the same problem about an unknown function f(r(X)) defined on the already estimated RIM $\mathbf{M}_{h,g}$ (9) whose solution can be taken as solution of the initial Problem. The function f(r(X)) can be written as

$$f(r(X)) = g_f(h(X)) \qquad (12)$$

in which

$$g_f(y) = f(g(y)) \tag{13}$$

is unknown function defined on the FS Y_h and, that, because of the relations (11) and (12), satisfies the proximity (4) and can be estimated from the feature sample $Z_{n,h}$ (5).

There exist independent solutions of each of the above two problems: the Input Manifold Estimation problem has been solved in [37] (this paper contains references to other related papers) and the Regression on features space estimation problem can be solved by various methods, see, for example, [4–9]. Therefore, it seems natural to solve these tasks consecutively, providing the solution (7) to the Regression on manifolds estimation problem; just such way was proposed in [38].

The trouble is that it is impossible constructing the consecutive and independent solutions to these problems: requirement (11) to the solution (h, g) to the Input Manifold Estimation problem, which is constructed on the first step, is formulated with use of unknown function f that is estimated only on next, Regression on features space estimation, step. Therefore, the Extended Regression on manifolds estimation problem consists in solving of following interrelated estimation tasks:

- Input Manifold Estimation problem, in which both the mapping (h, g) are constructed and their Jacobian matrices are estimated from the sample Z_n (1). These mappings determine the Feature space $Y_h = h(M)$ and the Recovered Input manifold $M_{h,g}$ (9) that satisfy the proximities (8), (10), (11);
- Regression on features space estimation problem, in which the mapping $g_f(y)$ (13) and its Jacobian matrix $J_{g,f}(y)$ defined on the FS Y_h are estimated from the feature sample $Z_{n,h}$ (5),

which should be solved in conjunction.

1.6   The Proposed Solution

We propose a new geometrically motivated approach to a solving of the Extended regression on manifolds estimation problem based on reducing this problem to certain Dimensionality reduction problem (namely, Tangent bundle manifold learning problem [39, 40]) for q-dimensional 'Regression manifold' $M(f) = \{(X, f(X))^T, X \in M\}$ embedded in input-output space R^{p+m}; hereinafter, the vectors are written as column-vectors, symbol ^{T} denotes transposition. All available information about unknown Regression manifold is contained in the dataset Z_n (1) sampled from this manifold.

The proposed approach gives also a new solution for common 'full-dimensional' Regression estimation problem in which an intrinsic dimension of the IM M is p. The similar approach has already been used in [41] for solving of common Regression problem and was results in new regression algorithm called Manifold Learning Regression; this algorithm allowed avoiding serious drawbacks that are intrinsic to many regression methods (kernel nonparametric regression, kriging, Gaussian processes) when they are applied to functions with strongly varying gradients [2, 8, 9].

This paper is organized as follows. Section 2 contains assumptions about the estimated objects and describes proposed approach to the Extended regression on manifolds estimation problem. The proposed solution is described in Sect. 3.

2   Extended Regression on Manifolds: A Problem Statement

2.1   Regression on Manifolds Estimation Problem: A Common Statement

Consider q-dimensional manifold

$$M = \{X = U(v) \in R^p : v \in V \subset R^q\} \tag{14}$$

covered by a single chart U and embedded in an ambient p-dimensional space R^p, q < p. The chart U is one-to-one mapping from open bounded space $V \subset R^q$ (called further Feature space, FS) to the manifold $M = U(V)$ with differentiable inverse map $U^{-1}: M \to V$. The intrinsic manifold dimension q is assumed to be known.

It is supposed that manifold **M** has positive condition number c(**M**) which is the number such that any point $X \in R^p$ distant from **M** by not more than $c^{-1}(M)$ has an unique projection onto the **M** [42], thus, no self-intersections, no 'short-circuit'; this means also that **M** has tubular neighborhood (ε-tube $B_\varepsilon(M)$) of radius $\varepsilon < c^{-1}(M)$.

Inverse function $h_u(X) = U^{-1}(X)$, whose values $v = h_u(X) \in V$ can be considered as low-dimensional coordinates on the manifold **M**, gives low-dimensional representations (features) $v = h_u(X)$ of high-dimensional manifold-valued data X. Note that pair (U, V) in the representation $M = U(V)$ (14) of the manifold **M** is determined up to arbitrary one-to-one mapping χ from the space R^q into itself: another pair (U*, V*), in which chart $U^*(v^*) = U(\chi^{-1}(v^*))$ is defined on Feature space $V^* = \chi(V)$, gives another representation $M = U^*(V^*)$ of the manifold **M**; this representation gives also another low-dimensional representations $v^* = h_u^*(X) = \chi(h_u(X))$ of manifold points $X \in M$.

If the mappings $h_u(X)$ and U(v) are differentiable and $J_{h,u}(X)$ and $J_u(v)$ are their $q \times p$ and $p \times q$ Jacobian matrices (covariant differentiation is used in $J_{h,u}(X)$ when $X \in M$), respectively, than q-dimensional linear space

$$L(X) = Span(J_u(h_u(X))) \tag{15}$$

in R^p is tangent space to the manifold **M** at the point $X \in M$; hereinafter, Span(H) is linear space spanned by columns of arbitrary matrix H. The tangent spaces can be considered as elements of the Grassmann manifold Grass(p, q) consisting of all q-dimensional linear subspaces in R^p. Note that tangent space L(X) (15) defined with use a concrete representation $M = U(V)$ (14) doesn't depend on chosen representation.

Let T = f(X) be differentiable mapping from input space R^p to output space R^m whose support Supp(f) contains ε-tube $B_\varepsilon(M)$, $\varepsilon < c^{-1}(M)$, of the manifold **M** called the Input manifold. Let $J_f(X) = \nabla_X f(X)$ be its $m \times p$ Jacobian matrix; covariant differentiation is used in $J_f(X)$ when we consider a restriction of the mapping f on the IM **M**. Common Regression on manifolds problem consists in estimating of the pair (f(X), $J_f(X)$) from given sample Z_n to provide proximities (2) and (3) for all points $X \in M$.

2.2   Regression on Manifolds Estimation: Related Tasks

Assume that the estimator $M_{h,g}$ (9) is consistent; it means that $d_H(M_{h,g}, M) \to 0$ with high probability as sample size $n \to \infty$; the term 'with high probability' means that the considered event holds with probability at least $(1 - C_\alpha /n^\alpha)$, where $\alpha \geq 1$ is an arbitrary number and C_α depends only on the number α (not on n). Such estimators exist, see, for example, the papers [37] in which $d_H(M_{h,g}, M) = O(n^{-2/(q+2)})$ [43]. Assume also that sample size n is sufficiently large to provide, with high probability, an inclusion $M_{h,g} \subset B_\varepsilon(M)$. Thus, the function f is defined on the RIM $M_{h,g}$.

Consider the function $f(r(X))$ defined on known (already estimated) manifold $M_{h,g}$ that has a representation (12), (13). Under conditions (11) (that implies condition (4)), this representation allows reducing the initial Regression on Manifolds Problem to the Regression on features space estimation problems about unknown function $g_f(y)$ defined on the FS Y_h. Therefore, the Regression on manifolds estimation problem requires solving of the following interrelated estimation tasks: given the sample Z_n (1),

- to construct a pair (h, g) consisting of Embedding and Recovering mappings, which determines the FS Y_h and the RIM $M_{h,g}$, satisfies the proximities (8) and (11), and provides 'differential proximity'

$$\nabla_X f(r(X)) \equiv J_{f,g}(h(X)) \times J_h(X) \approx \nabla_X f(X), \qquad (16)$$

where $m \times q$ and $q \times p$ matrices $J_{f,g}(y)$ and $J_h(X)$ are Jacobian matrices of the mappings $g_f(y)$ (13) and h(X) defined on the FS Y_h and IM M, respectively;
- to construct $q \times p$ and $p \times q$ matrices $G_h(X)$ and $G_g(y)$ defined on the IM M and FS Y_h, which estimate Jacobian matrix $J_h(X)$ and Jacobian matrix $J_g(y)$ of the mappings g, respectively, providing the proximities

$$G_h(X) \approx J_h(X), \qquad (17)$$

$$G_g(y) \approx J_g(y); \qquad (18)$$

- to estimate the mapping $g_f(y)$ (13) and its $m \times q$ Jacobian matrix $J_{f,g}(y)$ whose estimators $g_f^*(y)$ and $G_{f,g}(y)$ provide the proximities (6) and

$$G_{f,g}(y) \approx J_{f,g}(y). \qquad (19)$$

The solutions to above tasks determine estimator $f^*(X)$ (7) of the function f(X) and estimator

$$J^*(X) = G_{f,g}(h(X)) \times G_h(X) \qquad (20)$$

of the Gacobian matrix $J_f(X)$ that satisfy the required proximities (2) and (3).

Thus, the triples $(f^*(X), J^*(X), M_{h,g})$ and $(g_f^*(y), G_{f,g}(y), Y_h)$ give solution of the Regression on manifolds estimation problem.

2.3   Regression on Manifolds Estimation Problem: A Proposed Approach

We propose following approach to solution to the considered Extended regression on manifolds estimation problem that allows solving of all above interrelated tasks in conjunction.

An unknown function f defined on unknown IM **M** determines a smooth manifold

$$\mathbf{M}(f) = \{Z = F(X) \in R^r : X \in \mathbf{M} \subset R^p\} \tag{21}$$

embedded in an ambient input-output space R^r, $r = p + m$, in which

$$F : X \rightarrow Z = F(X) = \begin{pmatrix} x \\ f(x) \end{pmatrix} \in R^r \tag{22}$$

is the mapping defined on the ε-tube $B_\varepsilon(\mathbf{M})$ of the IM **M**.

The dataset $\mathbf{Z}_n$ (1) can be considered as a sample from the unknown manifold $\mathbf{M}(f)$. To distinguish between the IM **M** and introduced manifold $\mathbf{M}(f)$, the latter will be referred to as the Regression manifold (RM).

It follows from the representation (14) that the RM $\mathbf{M}(f)$ (21) can be written as

$$\mathbf{M}(f) = \{Z = F(U(v)) \in R^r : v \in \mathbf{V} \subset R^q\}$$

and, thus, is q-dimensional manifold embedded in ambient space R^r and covered by a single chart $F(U(v))$ defined on the FS **V**. The mapping $F(X)$ has $r \times p$ Jacobian matrix

$$J_F(X) = \begin{pmatrix} I_q \\ J_f(X) \end{pmatrix} \times \pi(X) \tag{23}$$

in which first multiplier ($r \times p$ matrix) is split into $p \times p$ unit matrix I_p and $m \times p$ Jacobian matrix $J_f(X)$ (covariant differentiation is used here). Second multiplier $\pi(X)$ in (23) is $p \times p$ projection matrix onto q-dimensional tangent space $L(X)$ (15) to the IM **M** at the point $X \in \mathbf{M}$. Thus, rank($J_F(X)$) = q and q-dimensional linear space

$$L_f(X) = \mathrm{Span}(J_F(X)) \in \mathrm{Grass}(r, q) \tag{24}$$

in R^r is tangent space to the RM $\mathbf{M}(f)$ at the point $F(X) \in \mathbf{M}(f)$, $X \in \mathbf{M}$.

First, we consider the Tangent bundle manifold learning (TBML) problem for the RM $\mathbf{M}(f)$ consisting in estimation of both the RM $\mathbf{M}(f)$ (21) and its tangent spaces $\{L_f(X), X \in \mathbf{M}\}$ (24) to the RM $\mathbf{M}(f)$ from the dataset $\mathbf{Z}_n$ sampled from the RM $\mathbf{M}(f)$.

We will use the previously proposed [39, 40] Grassmann&Stiefel Eigenmaps method (GSE) which gives a solution to this TBML problem and is described shortly in Sect. 3.1. The GSE-solution results in Embedding mapping $h_{GSE}(Z)$ from the $\mathbf{M}(f)$ to the Regression feature space (RFS) $\mathbf{Y}_{GSE} = h_{GSE}(\mathbf{M}(f)) \subset R^q$ and recovering mapping $g_{GSE}(y)$ from the RFS $\mathbf{Y}_{GSE}$ to the input-output space R^r, in such a way that

$$g_{GSE}(h_{GSE}(F(X))) \approx F(X) \tag{25}$$

for all points $Z = F(X) \in \mathbf{M}(f)$, or, the same, for all points $X \in \mathbf{M}$. In accordance with splitting of vector $Z \in \mathbf{R}^{p+m}$ into two vectors $Z_{in} \in \mathbf{R}^p$ and $Z_{out} \in \mathbf{R}^m$, the mapping $g_{GSE}(y)$ also can be split into two mappings $g_{GSE,in}$ and $g_{GSE,out}$:

$$g_{GSE}(y) = \begin{pmatrix} g_{GSE,in}(y) \\ g_{GSE,out}(y) \end{pmatrix} \tag{26}$$

from the RFS $\mathbf{Y}_{GSE}$ to the input space $\mathbf{R}^p$ and output space $\mathbf{R}^m$, respectively. Denote

$$h_{GSE,f}(X) = h_{GSE}(F(X)), \tag{27}$$

$$f_{GSE}(X) = g_{GSE,out}(h_{GSE,f}(X)), \tag{28}$$

then the approximate relations (25) can be written as

$$g_{GSE,in}(h_{GSE,f}(X)) \approx X, \tag{29}$$

$$f_{GSE}(X) \approx f(X), \tag{30}$$

for all $X \in \mathbf{M}$. Therefore,

- function $f_{GSE}(X)$ (28) accurately reconstructs (30) unknown function f,
- function $h_{GSE,f}(X)$ (27) maps the IM $\mathbf{M}$ into the RFS $\mathbf{Y}_{GSE} = h_{GSE,f}(\mathbf{M})$,
- q-dimensional manifold

$$\mathbf{M}_{GSE} = \{g_{GSE,in}(y) \in \mathbf{R}^p : y \in \mathbf{Y}_{GSE} \subset \mathbf{R}^q\}, \tag{31}$$

which is embedded in input space $\mathbf{R}^p$ and covered by a single chart $g_{GSE,in}(y)$ defined on the RFS $\mathbf{Y}_{GSE}$, is GSE-based estimator for the IM $\mathbf{M}$ and meets proximity

$$\mathbf{M}_{GSE} \approx \mathbf{M} \tag{32}$$

that follows from (29) and inequality $d_H(\mathbf{M}_{GSE}, \mathbf{M}) \leq \sup_{X \in \mathbf{M}} |g_{GSE,in}(h_{GSE,f}(X)) - X|$.

The GSE allows also constructing certain $m \times p$ matrix $G_{GSE,f}(X)$ that accurately approximates Jacobian $J_f(X)$ for all points $X \in \mathbf{M}$ and, thus, can be taken as estimator $J^*(X)$ that meets proximity (3). The first step is described in details in Sect. 3.2.

The triple $(f_{GSE}(X), G_{GSE,f}(X), \mathbf{M}_{GSE})$ accurately approximates the unknown triple $(f(X), J_f(X), \mathbf{M})$; the triple $(g_f(y), J_{f,g}(y), \mathbf{Y}_h)$ is also accurately estimated by certain GSE-based statistic $(g_{GSE,f}(y), G_{GSE,f,g}(y), \mathbf{Y}_{GSE})$. But these GSE-based triples cannot be considered as solution of the Regression on manifolds estimation problem because they depend on unknown function f.

Because of this, we consider these GSE-based triples as preliminary solutions based on which we construct in the second step the approximations for these triples that depend on the sample Z_n only. Thus, these approximations give final solution to the Extended regression on manifolds estimation problem.

The details of this second step are described in Sect. 3.3.

3 Extended Regression on Manifolds: A Solution

3.1 GSE-Solution to the Tangent Bundle of Regression Manifold Estimation

As was shown in [40], a minimization of reconstruction error $\delta(X)$ when solving Input Manifold Estimation problem implies certain additional requirement to the Embedding and Reconstruction mappings h and g: they should ensure not only Manifold proximity (8), (10), but also provide another property consisting in proximity between the tangent spaces to initial and recovered manifolds defined in terms of certain metric on the Grassmann manifold Grass(p, q).

In topology, a set composed of points of some manifold equipped by tangent spaces at these points is known as Tangent Bundle of the manifold. An amplification of the Manifold Estimation problem consisting in accurate reconstruction not only of manifold points but also the tangent spaces at these points can be referred to as the Tangent bundle manifold learning problem [39, 40].

The paper [40] contains theoretical justification of the need of solving the TBML for providing good generalization ability properties of solution to the Manifold Estimation problem and describes the GSE method that solves this amplified problem.

The GSE, which is being applied to a solving of the TBML for the RM $M(f)$, constructs an Embedding mapping $h_{GSE}(Z)$, which maps the RM $M(f)$ to the RFS Y_{GSE}, and a Recovering mapping $g_{GSE}(y)$ (26), which maps the RFS Y_{GSE} to the input-output space R^r, which ensure proximities (25) and

$$\nabla_X g_{GSE}(h_{GSE}(F(X))) \approx \nabla_X F(X),$$

covariant differentiation is used here. This equality can be written as

$$J_{GSE,g}(h_{GSE}(F(X)) \times J_{GSE,h,f}(X) \approx J_F(X) \tag{33}$$

in which $r \times q$ matrix $J_{GSE,g}(y)$ and $q \times p$ matrix $J_{GSE,h,f}(X)$ are Jacobian matrices of the mappings $g_{GSE}(y)$ and $h_{GSE}(F(X))$, respectively. Thus, the required proximities (16) are met. The GSE also constructs $r \times q$ matrix $G_{GSE,g}(y)$ that satisfies proximity

$$G_{GSE,g}(y) \approx J_{GSE,g}(y). \tag{34}$$

The $r \times q$ matrices $J_{GSE,g}(y)$ and $G_{GSE,g}(y)$ can be split into $p \times q$ matrices $J_{GSE,g,in}(y)$, $G_{GSE,in}$ and $m \times q$ matrices $J_{GSE,g,out}(y)$, $G_{GSE,out}$, respectively:

$$J_{GSE,g}(y) = \begin{pmatrix} J_{GSE,g,in}(y) \\ J_{GSE,g,out}(y) \end{pmatrix}, \quad G_{GSE,g}(y) = \begin{pmatrix} G_{GSE,in}(y) \\ G_{GSE,out}(y) \end{pmatrix}.$$

Using these representations and taking into account representation (23) and equalities (34), approximate equalities (33) can be written as

$$G_{GSE,in}(h_{GSE,f}(X)) \times J_{GSE,h,f}(X) \approx \pi(X), \tag{35}$$

$$G_{GSE,out}(h_{GSE,f}(X)) \times J_{GSE,h,f}(X) \approx J_f(X); \tag{36}$$

multiplier $\pi(X)$ is absent on the right side of Eq. (36) because Jacobian $J_f(X)$ has already been calculated with use covariant differentiation.

The q-dimensional linear space $L_{GSE}(X) = \mathrm{Span}(G_{GSE,in}(h_{GSE,f}(X)))$ is tangent space to the estimator $\mathbf{M}_{GSE}$ (31) of the IM $\mathbf{M}$ at the point $g_{GSE,in}(h_{GSE,f}(X)) \in \mathbf{M}_{GSE}$ that meets 'tangent' proximity $L_{GSE}(X) \approx L(X)$. Let $G_{GSE,in}(y) = G_{GSE,in,ort}(y) \times R(y)$ be QR-decomposition [45] of matrix $G_{GSE,in}(y)$ in which $G_{GSE,in,ort}$ is p × q orthogonal matrix and $R(y)$ is q × q nonsingular upper triangular matrix. Then p × p projection matrix

$$\pi_{GSE}(X) = G_{GSE,in,ort}(h_{GSE,f}(X)) \times G_{GSE,in,ort}^T(h_{GSE,f}(X)) \tag{37}$$

onto the tangent space $L_{GSE}(X)$ accurately approximates the projection matrix $\pi(X)$.

Considering the relations (35), in which projector $\pi(X)$ is replaced by the projector $\pi_{GSE}(X)$ (37), as regression equations about unknown matrix $J_{GSE,h,f}(X)$, standard Least-squares technique gives the q × p matrix

$$G_{GSE,h,f}(X) = G_{GSE,in}^-(h_{GSE,f}(X)) \times \pi_{GSE}(X), \tag{38}$$

which estimatess the Jacobian $J_{GSE,h,f}(X)$:

$$G_{GSE,h,f}(X) \approx J_{GSE,h,f}(X) \tag{39}$$

for all points $X \in \mathbf{M}$, here

$$G_{GSE,in}^-(h_{GSE,f}(X)) = (G_{GSE,in}^T(h_{GSE,f}(X)) \times G_{GSE,in}(h_{GSE,f}(X)))^{-1} \\ \times G_{GSE,in}^T(h_{GSE,f}(X)) \tag{40}$$

is q × p pseudoinverse Moore-Penrose matrix [45].

Finally, consider p × q matrix

$$G_{GSE,f}(X) = G_{GSE,out}(h_{GSE,f}(X)) \times G_{GSE,h,f}(X), \tag{41}$$

which, due to relations (36) and (39), meets required proximity

$$G_{GSE,f}(X) \approx J_f(X). \tag{42}$$

The constructed GSE-based triple $(f_{GSE}(X), G_{GSE,f}(X), M_{GSE})$ (28), (41), (31) meets all the required proximities (30), (42), (32). The mapping $g_{GSE,out}(y)$ (26) at the point $y = h_{GSE,f}(X) \in Y_{GSE}$ estimates the function $f(X)$ (28), (30) and meets relation (4) with $g_f(y) = f(g_{GSE,in}(y))$ (5). Therefore, the triple $(g_{GSE,out}(y), G_{GSE,out}(y), Y_{GSE})$ can be considered as solution to the Regression on features space estimation problem. But these triples cannot be regarded as a solution of the Regression on Manifolds Problem because depends on unknown function f.

By constructing, the GSE-based dependencies $h_{GSE,f}(X)$ and $g_{GSE}(y)$, as well as the estimators $G_{GSE,h,f}(X)$ (38) and $G_{GSE,g}(y)$ of their Jacobian matrices, are known at sample points $\{(X_i, y_{ih} = h_{GSE}(Z_i)), i = 1, 2, \ldots, n\}$. We construct approximations for these dependencies, which depend on the sample Z_n (1) only and meet also all the required proximities at Out-of-Sample points, with use proposed unified approach to solving of such regression problems with known Jacobian matrices at sample points called Known Jacobian Regression problems (KJR).

Next Sect. 3.2 contains some preliminaries (descriptions of used kernels, proposed solution of the KJR, choice of kernel bandwidths). The approximations of the GSE-based dependencies, which give final solution to the Regression on Manifolds estimation Problem, are constructed in Sect. 3.3.

3.2 Extended Regression on Manifolds: Preliminaries

Used Kernels. The GSE-based dependencies, which are defined either on the IM **M** or on the FS **Y** = h(**M**), are estimated by certain kernel estimators. In this Section, we describe the kernels used in estimation procedures. Hereinafter, by $\{\varepsilon_i\}$ we will denote algorithm parameters which are small positive numbers.

Kernel on Input manifold. Consider usually used kernel $K_E(X, X') = I\{|X' - X| < \varepsilon_1\}$ called 'Euclidean' kernel in which $I\{\cdot\}$ is indicator function.

Introduce a set $U_n(X, \varepsilon) \subset X_n$ consisting of sample points that belong to ε-ball in R^p centered at $X \in M$. An applying of the Principal Component Analysis (PCA) [46] to the set $U_n(X, \varepsilon_1)$ results in $p \times q$ orthogonal matrix $Q_{PCA}(X)$ whose columns are the PCA principal eigenvectors corresponding to the q largest PCA eigenvalues.

Consider q-dimensional linear spaces $L_{PCA}(X) = \text{Span}(Q_{PCA}(X))$ in the R^p considered as elements of the Grassmann manifold Grass(p, q) which, under certain conditions, satisfy proximities

$$L_{PCA}(X) \approx L(X) \text{ for all } X \in M \tag{43}$$

between the introduced spaces and tangent spaces $L(X)$ (15) to the IM **M**. Let

$$d_{BC}(L_{PCA}(X), L_{PCA}(X')) = \{1 - \text{Det}^2[Q_{PCA}^T(X) \times Q_{PCA}(X')]\}^{1/2},$$

$$K_{BC}(L_{PCA}(X), L_{PCA}(X')) = \text{Det}^2[Q_{PCA}^T(X) \times Q_{PCA}(X')],$$

be the Binet-Cauchy metric and Binet-Cauchy kernel, respectively, on the Grassmann manifold [47, 48]. Consider another data-based 'Grassmann' kernel

$$K_G(X, X') = I\{d_{BC}(L_{PCA}(X), L_{PCA}(X')) < \varepsilon_2\} \times K_{BC}(L_{PCA}(X), L_{PCA}(X'))$$

on the IM **M** and introduce a new 'aggregate' kernel

$$K(X, X') = K_E(X, X') \times K_G(X, X') \tag{44}$$

which reflects not only Euclidean nearness between X and X' but also nearness between the linear spaces $L_{PCA}(X)$ and $L_{PCA}(X')$, which, due the relations (43), results in nearness between the tangent spaces $L(X)$ and $L(X')$ to the IM **M**.

Kernel bandwidths. In what follows, we assume that the IM **M** is 'well-sampled' (it means that sample size n is large enough) to ensure a positive value of the qth eigenvalue in the PCA and provide proximities (43). To provide a trade-off between 'statistical error' in (43) depending on number of sample points in the set $U_n(X, \varepsilon_1)$ and 'curvature error' caused by deviation of the manifold-valued sample points from the 'assumed in the PCA' linear space, the ball radius ε_1 should tend to 0 as $n \to \infty$ with rate $O(n^{-1/(q+2)})$; this ensures the same order $O(n^{-1/(q+2)})$ for the PCA-error (with high probability) in (43) [49, 50]. A more accurate choice of the parameters ε_1 and ε_2 can be based on results of non-asymptotic analysis of the PCA estimators [51].

Kernel on Feature space. The GSE provides feature kernel $k_{GSE}(y, y')$ defined on the pairs $(y, y_{i,h})$, in which $y \in \mathbf{Y}_{GSE}$ and

$$y_{i,h} = h_{GSE}(Z_i) = h_{GSE,f}(X_i), \ i = 1, 2, \ldots, n, \tag{45}$$

are known sample features, that meets proximity

$$k_{GSE}(h_{GSE,f}(X), \ h_{GSE,f}(X_i)) \approx K(X, \ X_i)$$

for all $X \in \mathbf{M}$ and $i = 1, 2, \ldots, n$. Denote $k_{GSE}(y) = \sum\limits_{i=1}^{n} k_{GSE}(y, y_{i,h})$.

Estimation of Projectors onto Tangent Spaces. The $p \times p$ matrix

$$\pi_{PCA}(X) = Q_{PCA}(X) \times Q_{PCA}^T(X) \tag{46}$$

is projection matrix onto the linear space $L_{PCA}(X)$; it follows from (43) that $\pi_{PCA}(X)$ accurately approximates the projector $\pi(X)$ (15) and, thus, the projector $\pi_{GSE}(X)$ (40).

Regression problems with known Jacobians. Let $W = \Phi(S)$ be an unknown smooth mapping from its domain of definition $\mathbf{S} \subset \mathbf{R}^s$ to the $\mathbf{R}^t$. Let

$$\{(W_i = \Phi(S_i), \ J_{\Phi,i} = J_\Phi(S_i)), \ i = 1, 2, \ldots, n\} \tag{47}$$

be given sample consisting of known values of both the unknown function $\Phi(S)$ and its $t \times s$ Jacobian matrix $J_\Phi(S)$ at sample points $\{S_i \in \mathbf{S}\}$. The problem is to estimate $\Phi(S)$, $S \in \mathbf{S}$, from the sample (47).

Standard Kernel Nonparametric Regression (KNR) approach [52] to this problem is as follows. Let K(S, S′) be some chosen kernel on the domain of definition **S** reflecting nearness between the points S and S′. Then KNR-based estimator

$$\Phi_{KNR}(S) = \frac{1}{K(S)} \sum_{i=1}^{n} K(S,S_i) \times W_i, \tag{48}$$

where $K(S) = \sum_{j=1}^{n} K(S, S_j)$, minimizes the residual

$$\Delta_1(\Phi, S) = \sum_{i=1}^{n} K(S, S_i) \times |\Phi - \Phi(S_i)|^2.$$

Write the Taylor series expansion $\Phi(S) \approx \Phi(S′) + J_\Phi(S′) \times (S - S′)$ for close points S′, S ∈ **S** and consider the following 'second-order' residual function

$$\Delta_2(\Phi, S) = \sum_{i=1}^{n} K(S,S_i) \times |\Phi - W_i - J_{\Phi,i} \times (S - S_i)|^2 \tag{49}$$

depending on vector $\Phi \in R^t$. The vector

$$\Phi_{KJR}(S) = \Phi_{KNR}(S) + \frac{1}{K(S)} \sum_{i=1}^{n} K(S,S_i) \times J_{\Phi,i} \times (S - S_i), \tag{50}$$

called Known Jacobian Regression (KJR) estimator, minimizes over Φ residual (49).

The residual (49) and estimator (50) can be used only when values $J_{\Phi,i}$ of Jacobian matrices $J_\Phi(S_i)$ are known (or have been already estimated). Because of this, the considered problem is referred to as the Known Jacobian Regression problem.

If the kernel K(S, S′) equals 0 when $|S - S′| > \varepsilon$, where ε is small parameter, than the estimators (48) and (50) have the asymptotic expansions as $\varepsilon \to 0$:

$$\Phi_{KNR}(S) = \Phi(S) + \varepsilon \times \Phi_1(S) + o(\varepsilon), \tag{51}$$

$$\Phi_{KJR}(S) = \Phi(S) + \varepsilon^2 \times \Phi_2(S) + o(\varepsilon^2),$$

where $\Phi_1(S) = J_\Phi(S) \times \left[\frac{1}{\varepsilon \times K(S)} \sum_{i=1}^{n} K(S, S_i) \times (S_i - S)\right]$ and $\Phi_2(S)$ is t-dimensional vector whose j-th component $\Phi_{2,j}(S)$ equals to

$$\Phi_{2,j}(S) = \frac{1}{2\varepsilon^2 \times K(S)} \sum_{i=1}^{n} K(S, S_i) \times \left[(S_i - S)^T \times Hess_{\Phi,j}(S) \times (S_i - S)\right],$$

here $Hess_{\Phi,j}(S)$ is s × s Hessian matrix of j-th component $\Phi_j(S)$ of t-dimensional vector function $\Phi(S)$, j = 1, 2, ..., t.

Therefore, relations $|\Phi_{KNR}(S) - \Phi(S)| = O(\varepsilon)$ and $|\Phi_{KJR}(S) - \Phi(S)| = O(\varepsilon^2)$ hold under small ε. The best accuracy of estimator $\Phi_{KJR}(S)$ is due to the presence of second 'correction' summand in (50) which removes principal error term $\varepsilon \times \Phi_1(S)$ in asymptotic expansion (51) of the estimator $\Phi_{KNR}(S)$. This effect, in turn, is explained by used second-order residual (49) available under known Jacobians at sample points.

The $t \times s$ Jacobian matrix $J_\Phi(S)$ can be estimated by KNR-based statistic

$$J_{\Phi,\text{KNR}}(S) = \frac{1}{K(S)} \sum_{i=1}^{n} K(S,S_i) \times J_{\Phi,i} \qquad (52)$$

that satisfies proximity $J_{\Phi,\text{KNR}}(S) \approx J_\Phi(S)$.

3.3   Extended Regression on Manifolds Estimation: Final Solution

In this section, we construct the estimators of the GSE-based Embedding mapping $h_{\text{GSE,f}}(X)$ (27) and its Jacobian matrix $G_{\text{GSE,h,f}}(X)$ (38) (40), as well as the estimators of the GSE-based Recovering mapping $g_{\text{GSE}}(y)$ (26) and its Jacobian matrix $G_{\text{GSE,g}}(y)$.

Estimation of Embedding mapping and its Jacobian. Consider following Embedding regression estimation problem: to construct the Embedding mapping $h(X)$ which accurately approximates the mapping $h_{\text{GSE,f}}(X)$ (27):

$$h(X) \approx h_{\text{GSE,f}}(X), \ X \in \mathbf{M}, \qquad (53)$$

from known values $h_{\text{GSE,f}}(X_i) = h_{\text{GSE}}(Z_i)$ of the mapping $h_{\text{GSE,f}}(X)$ and known values $G_{\text{GSE,h}}(X_i)$ of its Jacobian matrix $G_{\text{GSE,h,f}}(X)$ (38), (40) at sample point $\{X_i\}$.

The proposed KJR- estimator (50) in the considered problem, in which the kernel (44) is used and projector $\pi_{\text{PCA}}(X)$ (46) replaces the projector π_{GSE} (40) in Jacobian matrix $G_{\text{GSE,h,f}}(X)$ (38), results in the estimator

$$h(X) = \frac{1}{K(X)} \sum_{i=1}^{n} K(X,X_i) \times \left\{ h_{\text{GSE}}(Z_i) + G_{\text{GSE,in}}^{-}(y_{ih}) \times \pi_{\text{PCA}}(X_i) \times (X - X_i) \right\}$$

$$(54)$$

of the Embedding mapping $h_{\text{GSE,f}}(X)$; the known sample features (45) are used. This estimator provides required proximity (53) and determines Feature space $\mathbf{Y}_h = h(\mathbf{M})$.

The matrix $G_{\text{GSE,h,f}}(X)$ (38) is estimated from its known values $\{G_{\text{GSE,h,f}}(X_i)\}$ at sample points with use KNR-estimator (48) and results in estimator

$$G_h(X) = \frac{1}{K(X)} \sum_{j=1}^{n} K(X,X_i) \times G_{\text{GSE,in}}^{-}(h(X_i)) \times \pi_{\text{PCA}}(X_i) \qquad (55)$$

that meets the required proximity (17).

Estimation of Recovering mapping and its Jacobian. The problem is to construct the recovering mapping

$$g(y) = \begin{pmatrix} g_{\text{in}}(y) \\ g_{\text{out}}(y) \end{pmatrix} \approx g_{\text{GSE}}(y) \qquad (56)$$

defined on the FS $\mathbf{Y}_h$ on the basis of known values $\{g_{\text{GSE}}(y_{i,h})\}$ and $\{G_{\text{GSE,g}}(y_{i,h})\}$ of the mapping $g_{\text{GSE}}(y)$ (26) and its Jacobian $G_{\text{GSE,g}}(y)$ at sample features (45).

The KJR- estimator (50), in which feature kernel is used, results in the estimators

$$g_{in}(y) = \frac{1}{k_{GSE}(y)} \sum_{i=1}^{n} k_{GSE}(y, y_{i,h}) \times \left\{ X_i + G_{GSE,in}(y_{i,h}) \times (y - y_{i,h}) \right\}, \quad (57)$$

$$g_{out}(y) = \frac{1}{k_{GSE}(y)} \sum_{i=1}^{n} k_{GSE}(y, y_{i,h}) \times \left\{ T_i + G_{GSE,out}(y_{i,h}) \times (y - y_{i,h}) \right\}, \quad (58)$$

that meets proximity (56).

The mappings (57) and (58) determine both the Input manifold estimator

$$\mathbf{M}_{in} = \left\{ g_{in}(y) \in R^p : y \in \mathbf{Y}_h \subset R^q \right\} \quad (59)$$

that, due to proximities (56) and (32), meets Manifold proximities

$$\mathbf{M}_{in} \approx \mathbf{M}_{GSE} \approx \mathbf{M}, \quad (60)$$

and the estimator

$$f_{out}(X) = g_{out}(h(X)), \quad (61)$$

that, due to the notations (28) and proximities (56), (53), and (30), meets proximities

$$f_{out}(X) \approx f_{GSE}(X) \approx f(X). \quad (62)$$

The Jacobian $G_{GSE,g}(y)$ is estimated from known values $\{G_{GSE,g}(y_{i,h})\}$ at feature sample points (45) with use KNR-estimator (48) that results in estimator

$$G_g(y) = \left(\frac{1}{k_{GSE}(y)} \sum_{i=1}^{n} k_{GSE}(y, y_{i,h}) \times G_{GSE,g}(y_{i,h}) \right) \quad (63)$$

which meets proximity $G_g(y) \approx G_{GSE,g}(y)$, and, hence, required proximity (18).

Jacobian of Unknown Mapping Estimation. The estimators $G_g(y)$ (63), $G_h(X)$ (55), and $h(X)$ (54) determine the estimator

$$G_f(X) = G_g(h(X)) \times G_h(X) \quad (64)$$

that meets required proximity

$$G_f(X) \approx G_{GSE,f}(X) \approx J_f(X). \quad (65)$$

Thus, the triple $(f_{out}(X), G_f(X), \mathbf{M}_{in})$ defined by formula (61), (64), and (59), meets the required proximities (62), (65), and (60). The triple $(g_{out}(y), G_g(y), \mathbf{Y}_h = h(\mathbf{M}))$ defined by formula (58) and (63), gives the solution to the Regression on feature space estimation problem.

4 Conclusion

The paper considers Extended Regression on manifold estimation problem in which a few interrelated problems are solved in conjunction. First of them is common Regression on manifold problem in which unknown m-dimensional function f(X) of p-dimensional manifold-valued inputs X, whose values lie on unknown Input manifold M of lower dimensionality q < p embedded in an ambient high-dimensional space R^p, and its Jacobian matrix $J_f(X)$, are estimated from given sample Z_n consisting of 'input-output' pairs. Input manifold estimation problem is to estimate unknown domain of definition M of function f; this problem includes a constructing of mapping h that transforms M to q-dimensional Feature space $Y_h = h(M)$ and satisfies certain conditions. Regression on feature space estimation problem consists in estimating of both a specified unknown function $g_f(y)$, which depends on low-dimensional features y = h(X) $\in Y_h$ and satisfies the condition $g_f(h(X)) \approx f(X)$, and its Jacobian $J_{g,f}(y)$.

The proposed construction reduces initial 'high-dimensional' estimation problem about f(X), X $\in$ M $\subset$ R^p, to similar 'low-dimensional' estimation problem about $g_f(y)$, y $\in Y_h \subset R^q$, and allows avoiding curse of dimensionality phenomenon.

We propose a new geometrically motivated method for solving the Extended regression on manifolds estimation problem which is based on reducing this problem to certain Dimensionality reduction problem (namely, Tangent bundle manifold learning) for an unknown q-dimensional Regression manifold $M(f) = \{(X, f(X))^T, X \in M\}$ embedded in input-output space R^{p+m}; all available information about Regression manifold is contained in the dataset Z_n sampled from its manifold.

The previously proposed Grassmann&Stiefel Eigenmaps method, which is used for solving of this Tangent bundle manifold learning problem, and proposed estimators in specific regression problem, in which Jacobian matrices of the estimated mapping are known at the sample points, allowed estimating of unknown triples (f(X), $J_f(X)$, M) and ($g_f(y)$, $J_{g,f}(y)$, Y_h) from the given sample.

The proposed solution gives also a new solution to common 'full-dimensional' Regression estimation problem in which an intrinsic dimension of the IM M is p.

Acknowledgments. The study was performed in the IITP RAS exclusively by the grant from the Russian Science Foundation (project № 14-50-00150).

References

1. Vapnik, V.: Statistical Learning Theory. John Wiley, New-York (1998)
2. Hastie, T., Tibshirani, R., Friedman, J.: The Elements of Statistical Learning: Data Mining, Inference, and Prediction, 2nd edn. Springer, New York (2009)
3. James, G., Witten, D., Hastie, T., Tibshirani, R.: An Introduction to Statistical Learning with Applications in R. Springer Texts in Statistics. Springer, New-York (2013)
4. Bishop, C.M.: Pattern Recognition and Machine Learning. Springer, Heidelberg (2007)
5. Deng, L., Yu, D.: Deep Learning: Methods and Applications. NOW Publishers, Boston (2014)
6. Breiman, L.: Random forests. Mach. Learn. **45**(1), 5–32 (2001)

7. Friedman, J.H.: Greedy function approximation: a gradient boosting machine. Ann. Stat. **29** (5), 1189–1232 (2001)
8. Rasmussen, C.E., Williams, C.: Gaussian Processes for Machine Learning. MIT Press, Cambridge (2006)
9. Loader, C.: Local Regression and Likelihood. Springer, New York (1999)
10. Wang, G.G., Shan, S.: Review of metamodeling techniques in support of engineering design optimization. J. Mech. Des. **129**(3), 370–381 (2007)
11. Forrester, A.I.J., Sobester, A., Keane, A.J.: Engineering Design via Surrogate Modelling: A Practical Guide. Wiley, New-York (2008)
12. Kuleshov, A.P., Bernstein, A.V.: Cognitive technologies in adaptive models of complex plants. Inf. Control Probl. Manuf. **13**(1), 1441–1452 (2009)
13. Stone, C.J.: Optimal rates of convergence for nonparametric estimators. Ann. Stat. **8**, 1348–1360 (1980)
14. Stone, C.J.: Optimal global rates of convergence for nonparametric regression. Ann. Stat. **10**, 1040–1053 (1982)
15. Rajaram, D., Pant, R.S.: An improved methodology for airfoil shape optimization using surrogate based design optimization. In: Rodrigues, H., et al. (eds.) Engineering Optimization IV, pp. 147–152. CRC Press, Taylor & Francis Group, London (2015)
16. Bernstein, A., Kuleshov, A., Sviridenko, Y., Vyshinsky, V.: Fast aerodynamic model for design technology. In: Proceedings of West-East High Speed Flow Field Conference (WEHSFF-2007), Moscow, Russia (2007). http://wehsff.imamod.ru/pages/s7.htm
17. Zhu, F., Qin, N., Burnaev, E.V., Bernstein, A.V., Chernova, S.S.: Comparison of three geometric parameterization methods and their effect on aerodynamic optimization. In: Poloni, C. et al. (eds.) Eurogen 2011, Optimization and Control with Applications to Industrial and Societal Problems International Conference on Proceedings - Evolutionary and Deterministic Methods for Design, pp. 758–772. Sira, Capua, Italy (2011)
18. Seung, H.S., Lee, D.D.: The manifold ways of perception. Science **290**(5500), 2268–2269 (2000)
19. Levina, E., Bickel, P.J.: Maximum likelihood estimation of intrinsic dimension. In: Saul, L., Weiss, Y., Bottou, L. (eds.) Advances in Neural Information Processing Systems, vol. 17, pp. 777–784. MIT Press, Cambridge (2005)
20. Fan, M., Qiao, H., Zhang, B.: Intrinsic dimension estimation of manifolds by incising balls. Pattern Recogn. **42**, 780–787 (2009)
21. Fan, M., Gu, N., Qiao, H., Zhang, B.: Intrinsic dimension estimation of data by principal component analysis. In: arXiv:1002.2050v1 [cs.CV], 10 Feb 2010, pp. 1–8 (2010)
22. Rozza, A., Lombardi, G., Rosa, M., Casiraghi, E., Campadelli, P.: IDEA: intrinsic dimension estimation algorithm. In: Maino, G., Foresti, G.L. (eds.) ICIAP 2011, Part I. LNCS, vol. 6978, pp. 433–442. Springer, Heidelberg (2011)
23. Huo, X., Ni, X., Smith, A.K.: Survey of manifold-based learning methods. In: Liao, T.W., Triantaphyllou, E. (eds.) Recent Advances in Data Mining of Enterprise Data, pp. 691–745. World Scientific, Singapore (2007)
24. Ma, Y., Fu, Y. (eds.): Manifold Learning Theory and Applications. CRC Press, London (2011)
25. Bernstein, A., Kuleshov, A.: Low-dimensional data representation in data analysis. In: El Gayar, N., Schwenker, F., Suen, C. (eds.) ANNPR 2014. LNCS, vol. 8774, pp. 47–58. Springer, Heidelberg (2014)
26. Kuleshov, A., Bernstein, A.: Manifold learning in data mining tasks. In: Perner, P. (ed.) MLDM 2014. LNCS, vol. 8556, pp. 119–133. Springer, Heidelberg (2014)
27. Pelletier, B.: Nonparametric regression estimation on closed Riemannian manifolds. J. Nonparametric Stat. **18**(1), 57–67 (2006)

28. Loubes, J.-M., Pelletier, B.: A kernel-based classifier on a riemannian manifold. Statistics and Decisions **26**(1), 35–51 (2008). Verlag, Oldenbourg

29. Bickel, P., Li, B.: Local polynomial regression on unknown manifolds. IMS Lecture notes - Monograph Series, vol. 54 'Complex Datasets and Inverse Problems: Tomography, Networks and Beyond,' pp. 177–186 (2007)

30. Aswani, A., Bickel, P., Tomlin, C.: Regression on manifolds: Estimation of the exterior derivative. Ann. Stat. **39**(1), 48–81 (2011)

31. Cheng, M.-Y., Wu, H.-T.: Local linear regression on manifolds and its geometric interpretation. J. Am. Stat. Assoc. **108**(504), 1421–1434 (2013)

32. Yang, Y., Dunson, D.B.: Bayesian manifold regression. In: arXiv:1305.0167v2 [math.ST], 16 June 2014, pp. 1–40 (2014)

33. Guhaniyogi, R., Dunson, D.B.: Compressed gaussian process. In: arXiv:1406.1916v1 [stat. ML], 7 June 2014, pp. 1–29 (2014)

34. Fletcher, P.T.: Geodesic regression on Riemannian manifolds. In: Proceedings of International Workshop on Mathematical Foundations of Computational Anatomy (MFCA), pp. 75–86 (2011)

35. Hinkle, J., Muralidharan, P., Fletcher, P.T.: Polynomial regression on riemannian manifolds. In: arXiv:1201.2395v2 [math.ST], 1 Mar 2012, pp. 1–14 (2012)

36. Steinke, F., Hein, M., Schölkopf, B.: Nonparametric regression between general riemannian manifolds. SIAM J. Imaging Sci. **3**(3), 527–563 (2010)

37. Bernstein, A.V., Kuleshov, A.P.: Data-based manifold reconstruction via tangent bundle manifold learning. In: ICML-2014, Topological Methods for Machine Learning Workshop, Beijing, 25 June 2014. http://topology.cs.wisc.edu/KuleshovBernstein.pdf (2014)

38. Kuleshov, A.P., Bernstein, A.V.: Cognitive technologies in adaptive models of complex plants. Inf. Control Probl. Manuf. **13**(1), 1441–1452 (2009)

39. Bernstein, A.V., Kuleshov, A.P.: Tangent bundle manifold learning via Grassmann&Stiefel Eigenmaps. In: arXiv:1212.6031v1 [cs.LG], December 2012, pp. 1–25 (2012)

40. Bernstein, A.V., Kuleshov, A.P.: Manifold learning: generalizing ability and tangent proximity. Int. J. Softw. Inf. **7**(3), 359–390 (2013)

41. Bernstein, A., Kuleshov, A., Yanovich, Y.: Manifold Learning in Regression Tasks. In: Gammerman, A., Vovk, V., Papadopoulos, H. (eds.) SLDS 2015. LNCS, vol. 9047, pp. 414–423. Springer, Heidelberg (2015)

42. Niyogi, P., Smale, S., Weinberger, S.: Finding the homology of submanifolds with high confidence from random samples. Discrete Comput. Geom. **39**, 419–441 (2008)

43. Kuleshov, A., Bernstein, A., Yanovich, Yu.: Asymptotically optimal method in Manifold estimation. In: Márkus, L., Prokaj, V. (eds.) Abstracts of the XXIX-th European Meeting of Statisticians, 20–25 July 2013, Budapest, p. 325 (2013)

44. Genovese, C.R., Perone-Pacifico, M., Verdinelli, I., Wasserman, L.: Minimax manifold estimation. J. Mach. Learn. Res. **13**, 1263–1291 (2012)

45. Golub, G.H., Van Loan, C.F.: Matrix Computation, 3rd edn. Johns Hopkins University Press, Baltimore, MD (1996)

46. Jollie, T.: Principal Component Analysis. Springer, New-York (2002)

47. Hamm, J., Lee, D.D.: Grassmann discriminant analysis: a unifying view on subspace-based learning. In: Proceedings of the 25th International Conference on Machine Learning (ICML 2008), pp. 376–383 (2008)

48. Wolf, L., Shashua, A.: Learning over sets using kernel principal angles. J. Mach. Learn. Res. **4**, 913–931 (2003)

49. Singer, A., Wu, H.-T.: Vector diffusion maps and the connection laplacian. Commun. Pure Appl. Math. **65**(8), 1067–1144 (2012)

50. Tyagi, H., Vural, E., Frossard, P.: Tangent space estimation for smooth embeddings of Riemannian manifold. In: arXiv:1208.1065v2 [stat.CO], 17 May 2013, pp. 1–35 (2013)
51. Kaslovsky, D.N., Meyer, F.G.: Non-asymptotic analysis of tangent space perturbation. Inf. Inf. J. IMA **3**(2), 134–187 (2014)
52. Wasserman, L.: All of Nonparametric Statistics. Springer Texts in Statistics, Berlin (2007)

Author Index

Printed in the United States
By Bookmasters

Printed in the United States
By Bookmasters